Atlas of Hybrid Imaging
Sectional Anatomy for PET/CT, PET/MRI and SPECT/CT
Vol. 2: Thorax Abdomen and Pelvis

Atlas of Hybrid Imaging
Sectional Anatomy for PET/CT, PET/MRI and SPECT/CT
Vol. 2: Thorax Abdomen and Pelvis

Mario Leporace
Department of Nuclear Medicine and Theranostics, Cosenza Hospital, Italy

Ferdinando Calabria
Department of Nuclear Medicine and Theranostics, Cosenza Hospital, Italy

Eugenio Gaudio
Department of Human Anatomy, "La Sapienza" University, Rome, Italy

Orazio Schillaci
Department of Biomedicine and Prevention, "Tor Vergata" University, Rome, Italy

Alfonso Ciaccio
Department of Nuclear Medicine and Theranostics, Cosenza Hospital, Italy

Antonio Bagnato
Department of Nuclear Medicine and Theranostics, Cosenza Hospital, Italy

Academic Press is an imprint of Elsevier
125 London Wall, London EC2Y 5AS, United Kingdom
525 B Street, Suite 1650, San Diego, CA 92101, United States
50 Hampshire Street, 5th Floor, Cambridge, MA 02139, United States
The Boulevard, Langford Lane, Kidlington, Oxford OX5 1GB, United Kingdom

Copyright © 2023 Elsevier Inc. All rights reserved.

No part of this publication may be reproduced or transmitted in any form or by any means, electronic or mechanical, including photocopying, recording, or any information storage and retrieval system, without permission in writing from the publisher. Details on how to seek permission, further information about the Publisher's permissions policies and our arrangements with organizations such as the Copyright Clearance Center and the Copyright Licensing Agency, can be found at our website: www.elsevier.com/permissions.

This book and the individual contributions contained in it are protected under copyright by the Publisher (other than as may be noted herein).

Notices

Knowledge and best practice in this field are constantly changing. As new research and experience broaden our understanding, changes in research methods, professional practices, or medical treatment may become necessary.

Practitioners and researchers must always rely on their own experience and knowledge in evaluating and using any information, methods, compounds, or experiments described herein. In using such information or methods they should be mindful of their own safety and the safety of others, including parties for whom they have a professional responsibility.

To the fullest extent of the law, neither the Publisher nor the authors, contributors, or editors, assume any liability for any injury and/or damage to persons or property as a matter of products liability, negligence or otherwise, or from any use or operation of any methods, products, instructions, or ideas contained in the material herein.

ISBN: 978-0-443-18733-9

For information on all Academic Press publications visit our website at https://www.elsevier.com/books-and-journals

Publisher: Stacy Masucci
Acquisitions Editor: Katie Chan
Editorial Project Manager: Sam W. Young
Production Project Manager: Omer Mukthar
Cover Designer: Christian J. Bilbow

Typeset by TNQ Technologies

Printed in the United States of America
Last digit is the print number: 9 8 7 6 5 4 3 2 1

To my wife Fedora, my great love, and Luigivittorio   ML
and Niccolò, my beloved children.

To Giuliana, my best friend and one true love, and   FC
to Vittoria and Francesca Junia, the sweetest things
of our life.

To Ida, my beloved wife.   EG

To Nicoletta, Maria Beatrice, and Agnese Felicia.   OS

To my father Severino, better than a hundred teachers.   AC

To Mariella.   AB

Contents

Preface   ix
Acknowledgments   xi

1. Thorax

Introduction: 3D-CT volume rendering of
anatomy   2
1.1 Lung PET/CT   6
    1.1.1 Axial, sagittal, coronal (lobes and
fissures)   6
    1.1.2 Axial (bronchopulmonary segments)   10
    1.1.3 Sagittal (bronchopulmonary
segments)   54
    1.1.4 Coronal (bronchopulmonary
segments)   74
1.2 Mediastinum PET/CT   94
    1.2.1 Anatomy   94
    1.2.2 Axial   96
1.3 Clinical cases, tricks, and pitfalls   128
    1.3.1 ^{18}F-FDG   128
    1.3.2 ^{18}F-DOPA   145
    1.3.3 ^{68}Ga-DOTATOC   146
    1.3.4 ^{18}F-choline   147
    1.3.5 ^{18}F-NaF   150
    1.3.6 ^{131}I   151
    1.3.7 ^{99m}Tc-MDP   152
References   154

2. Abdomen and pelvis

Introduction: 3D-CT volume rendering of
anatomy   158
2.1 General anatomy PET/CT   170
    2.1.1 Axial   170
    2.1.2 Sagittal   226
    2.1.3 Coronal   242
2.2 Liver PET/CT   256
    2.2.1 Axial anatomy   256
2.3 Peritoneum and retroperitoneum
PET/CT   272
    2.3.1 Peritoneum   272
    2.3.2 Retroperitoneum   286
2.4 Pelvis and perineum PET/MRI   296
    2.4.1 Female pelvis   296
    2.4.2 Male pelvis   306
2.5 Clinical cases, tricks, and
pitfalls   316
    2.5.1 ^{18}F-FDG   316
    2.5.2 ^{68}Ga-DOTATOC   331
    2.5.3 ^{18}F-Choline   333
    2.5.4 ^{68}Ga-PSMA   336
    2.5.5 ^{99m}Tc   337
References   337

Index   341

Preface

The advent of hybrid scanners in nuclear medicine has considerably improved quality of the discipline.

To date, SPECT/CT and PET/CT play a crucial role in diagnosis, treatment planning, and assessment of response to therapy in oncology, with novel applications in neurology, cardiology, and infectious diseases. More recently, PET/MRI has considerably enlarged the panorama of hybrid imaging and is opening new challenges in neurooncology, oncology, and neurodegenerative disorders.

In fact, the high sensitivity provided by nuclear medicine imaging finds a valid counterpart in the better specificity provided by the *low-dose* CT, generally associated with PET, and, when possible, SPECT scans.

The added value of this coregistered *low-dose* CT is defined by a better diagnostic accuracy due to higher specificity, allowing adequate anatomical localization of pathologic functional findings and accurate depiction of false-positive or false-negative cases that can occur in clinical practice with all radiopharmaceuticals.

All PET scans in oncology are coregistered with a *low-dose* CT for attenuation correction and anatomical landmarks. The CT component of a PET/CT scan is also an authentic *trait d'union* between nuclear medicine imaging and contrast enhanced CT and MRI, being an anatomical basis for comparison of functional imaging with advanced morphological imaging. This feature is of the utmost importance; in fact, though not accurate as *full-dose* contrast-enhanced CT, the *low-dose* CT of PET/CT offers significant anatomical information (*i.e.*, on the lungs, bones, and soft tissues) strengthening confidence in diagnosis and helping nuclear physicians to compose more exhaustive medical reports.

These characteristics will have a significantly higher impact in PET/MRI realm, considering the large availability of sequences, the optimal power resolution limit of MRI, and correlative advanced studies of molecular imaging as Diffusion-Weighted Imaging or MR Spectroscopy.

Therefore, the challenge for nuclear physicians in the rising era of hybrid imaging is due to the following:

- the definitive transition from 2D to 3D medical images;
- accurate knowledge of anatomical landmarks in multiplanar hybrid views.

The *2.0 nuclear medicine* should be aimed to improve the quality of postprocessing and reports in order to optimize the dialogue with radiologists as well as oncologists and clinicians of diverse specialties.

It is also necessary to state that the versatility of ^{18}F-FDG (the *miliar stone* among PET tracers), and the rapid development of a large amount of PET tracers, enlarged PET molecular imaging to new fields of interest in neurology, cardiology, infectious diseases, and neurooncology, while new molecular frontiers are under investigation.

Some of these new molecular probes, ^{18}F—NaF, amyloid tracers, amino acid radiopharmaceuticals such as ^{18}F-FET, ^{18}F-DOPA, radiolabeled choline, and ^{68}Ga-labeled somatostatin analogues, significantly expanded the panorama of molecular imaging with PET.

Finally, the advent of *theragnostics*, a new discipline where *diagnosis* and *therapy* are strictly linked and modulated by radiopharmaceuticals and PET imaging, imposes to improve the knowledge of sectional anatomy for an optimal assessment of response to therapy.

It is worth to mention that *hybrid, molecular* imaging is the pillar of the new era of nuclear medicine, an eclectic discipline which is skin changing and evolving as leading medical specialty.

Several atlas are already available for intepretation of CT and MRI, but these volumes are generally aimed to radiologists. In our opinion of nuclear physicians and radiologists with expertise on hybrid imaging, a volume focused on residents in nuclear medicine and/or radiology or young specialists is still needed, as a *daily guide* to medical reports; however, it could easily be useful for nuclear physicians and/or radiologists aiming to improve experience in hybrid imaging or for specialists interested in diagnostic imaging (radiotherapists, cardiologists, neurologists, *etc.*).

The aim of our book is to give to young nuclear physicians and radiologists a rapid, concise guide to radiological anatomy as support for nuclear medicine findings, with emphasis on the role of coregistered CT and MRI in improving

x Preface

diagnosis and correctly detect false-positive and false-negative cases. Therefore, each chapter is focused on a specific segment of the human body, with three-dimensional introductive views followed by commented *low-dose* CT views in every plane, to depict sectional anatomy. Terminology of anatomical landmarks is preferentially in accordance with the International Anatomical Terminology, following guidelines of the Federative International Programme for Anatomical Terminology (FIPAT) of the International Federation of Associations of Anatomists (IFAA), with the addendum, when proper, of a correlative *radiological* synonym in routine practice.

All *low-dose* CT views are associated with corresponding PET/CT views and, as standard of quality imaging, contrast-enhanced CT or MRI. CT, MRI, and PET/CT images are also provided, when needed, by brief comments on their features, especially on the visual and/or semiquantitative assessment of tracer uptake in tissues, in order to introduce the reader to the analysis of fused PET/CT images, in color scale.

This volume provides a complete coverage of CT sectional anatomy of the brain, neck, thorax, abdomen, and pelvis, while special chapters are focused on the musculoskeletal system, cardiac imaging, and lymph nodes. PET/MRI schemes are also provided for some peculiar anatomical districts as the brain, neck, and pelvis.

Anatomical terms and descriptions are prevalently based on *Terminologia Anatomica*.

Moreover, only literature data explain the most commonly used PET tracers, depicting their peculiar molecular pathways, the physiological distribution, common pathologic findings, and diagnostic pitfalls. The currently available atlas of imaging, only focused on anatomy, does not address these contents.

At the end of each chapter, a special paragraph shows and discusses clinical cases, pitfalls, and anatomical variants, in order to explain peculiarities, intrinsic properties, concordance, or mismatches between nuclear medicine findings and CT or MRI, as well as teaching points to explain radiological imaging criteria. Clinical cases and pitfalls are referred to several PET and SPECT radiopharmaceuticals, in order to show and discuss peculiar conditions linked to specific tracers.

Essential references are reported in order to allow the readers to refer back to any source that can be linked to covered topics.

We hope to have centered our ambitious scope: to offer to nuclear physicians and colleagues a rapid, easy to consult, didactic guide toward hybrid, molecular imaging.

Mario Leporace
Ferdinando Calabria

Acknowledgments

We wish to express very great appreciation to our colleagues for their recommendations, enthusiastic encouragement, and help in difficult situations: Rosanna Tavolaro, Maria Toteda, Stefania Cardei, and Antonio Lanzillotta.

A debt of gratitude is owed to our *Roman colleagues* for their valuable suggestions: Armando Mancinelli and Mauro Di Roma.

A special thanks to Professor Antonio Cerasa: we carefully followed your experience!

We would like to particularly extend thanks to Daniela de Silva and technicians radiologists, and nurses of our department for their work and patients care.

Humbly, we are grateful to our professors and to the pioneers of our discipline for letting us see further by standing on their shoulders.

Chapter 1

Thorax

Abstract

Evaluation of the chest implies accurate knowledge of the lungs, muscles, airways, bone, and vascular structures. The low-dose CT of nuclear imaging allows recognition of anatomy and pathologic findings, essential for the optimal hybrid imaging report, using adequate CT window level. This chapter presents low-dose CT views (with corresponding ^{18}F-FDG PET/CT and contrast-enhanced CT) of the thorax with multiplanar pulmonary and mediastinal details, aiming to sample all lymph nodal, pulmonary, or vascular anatomical landmarks for nuclear physicians. Three-dimensional CT views of normal anatomy of lungs and mediastinum are showed, with a holistic approach. Several PET/CT and SPECT/CT clinical cases, tricks, and pitfalls are discussed, on different tracers: ^{18}F-FDG, ^{18}F-DOPA, ^{68}Ga-DOTATOC, ^{18}F-NaF, ^{18}F-choline, ^{131}I, and ^{99m}Tc-MDP. All tracer-related findings are reported with emphasis on the added role of PET imaging and adequate knowledge of CT in reaching final diagnosis.

Keywords: ^{18}F-choline; ^{18}F-DOPA; ^{18}F-FDG; ^{68}Ga-DOTATOC; CT; Lung cancer; Medullary carcinoma; PET/CT; Pulmonary node; Thorax

Introduction: 3D-CT volume rendering of anatomy

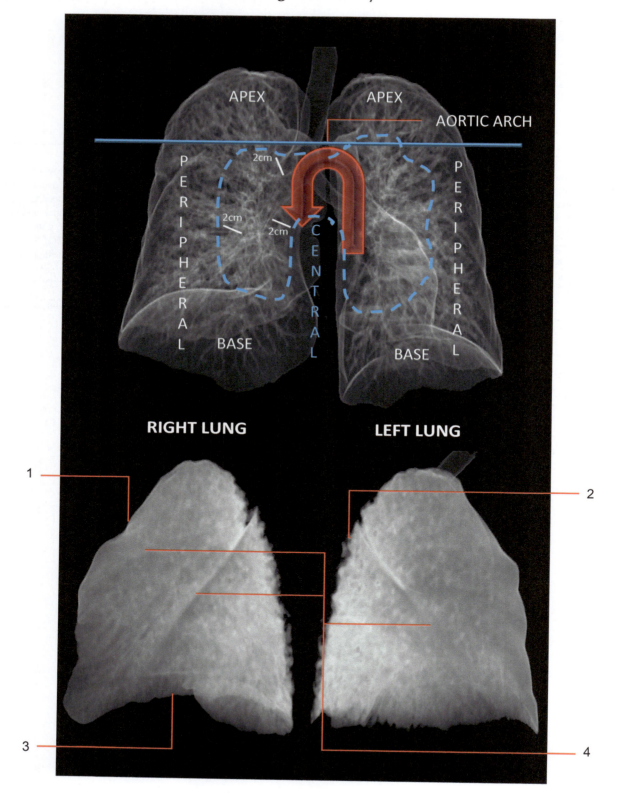

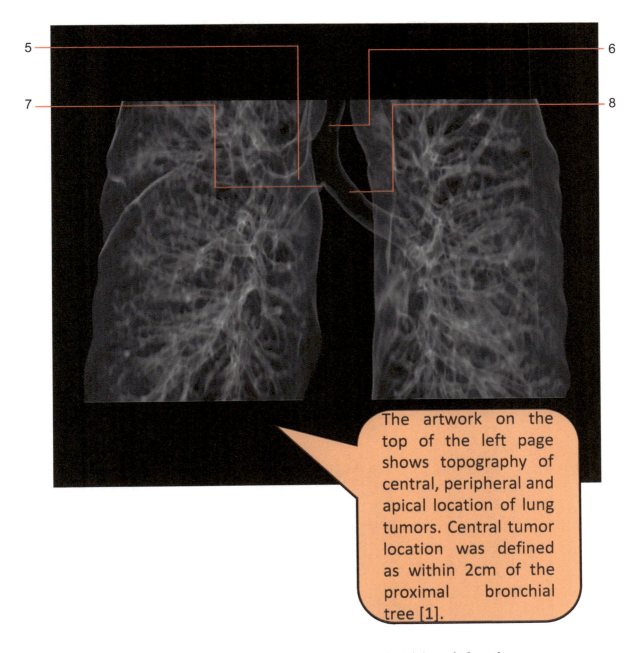

The artwork on the top of the left page shows topography of central, peripheral and apical location of lung tumors. Central tumor location was defined as within 2cm of the proximal bronchial tree [1].

1 Anterior border of the lung
2 Posterior border of the lung
3 Inferior border and diaphragmatic surface of the lung
4 Fissures (obliques/horizontal)
5 Right main bronchus
6 Trachea (thoracic part)
7 Carina of trachea
8 Left main bronchus

See Ref [1]

4   Atlas of Hybrid Imaging

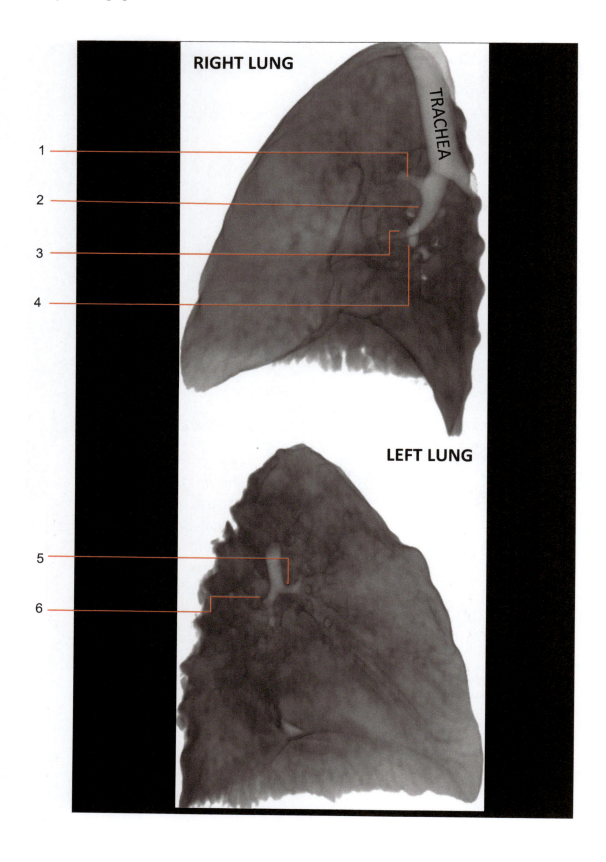

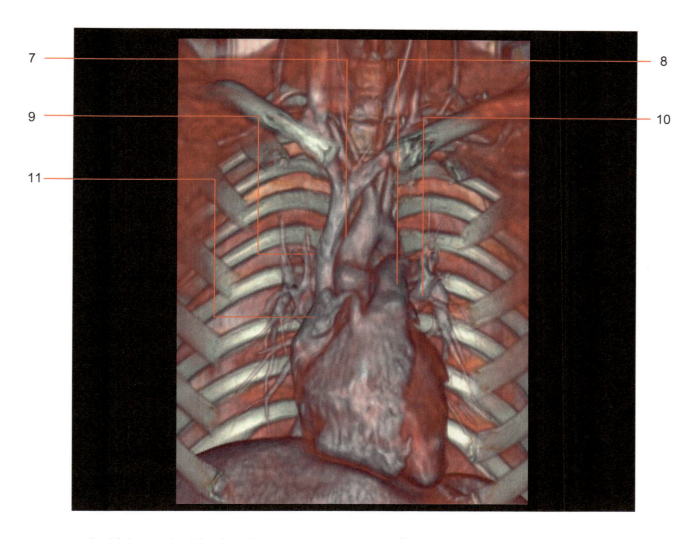

1 Right superior lobar bronchus
2 Intermediate bronchus
3 Middle lobar bronchus
4 Right inferior lobar bronchus
5 Left superior lobar bronchus
6 Left inferior lobar bronchus
7 Ascending aorta
8 Pulmonary trunk
9 Superior vena cava
10 Left pulmonary vein
11 Right auricle of the heart

1.1 Lung PET/CT

1.1.1 Axial, sagittal, coronal (lobes and fissures)

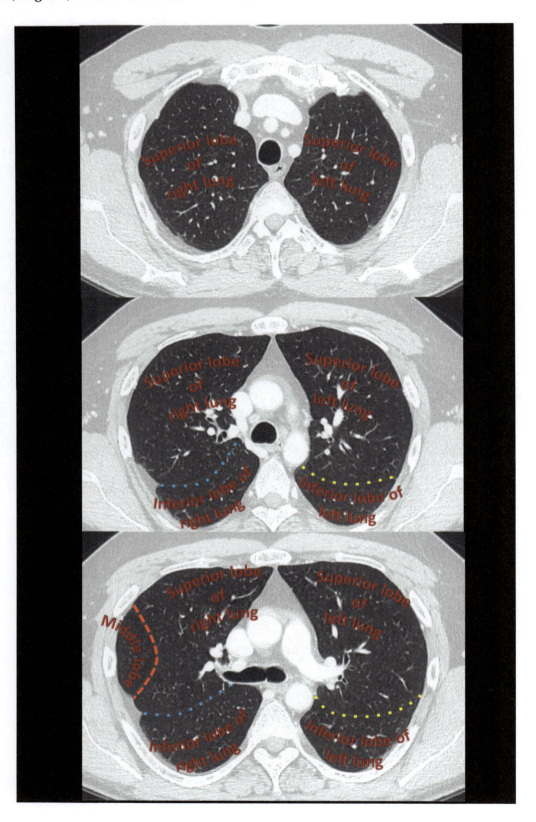

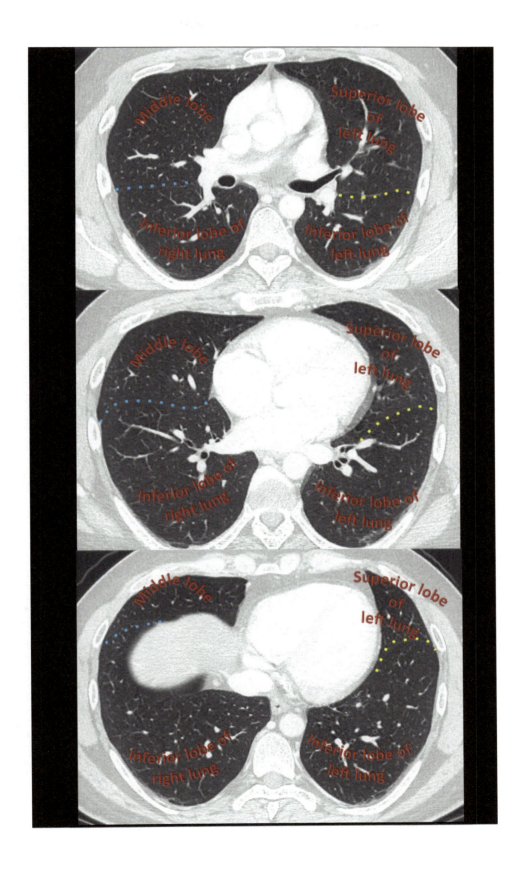

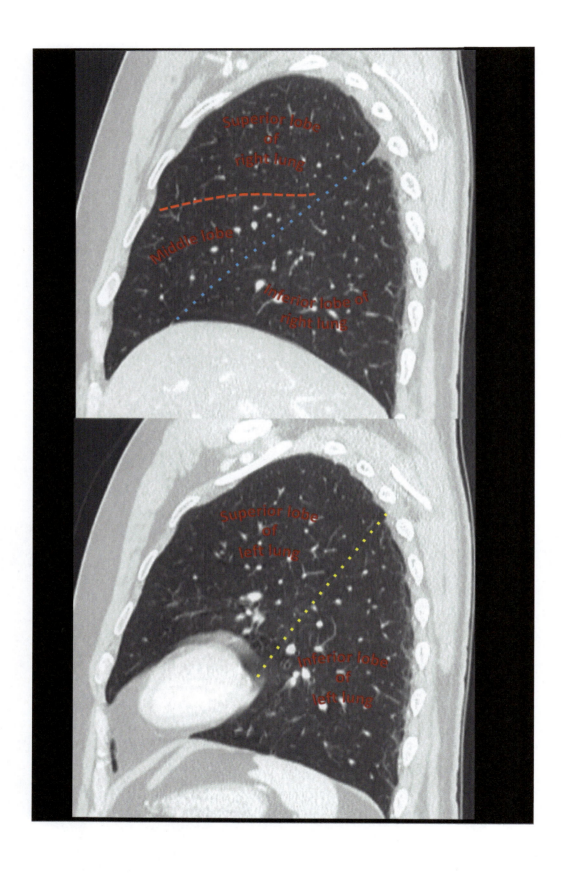

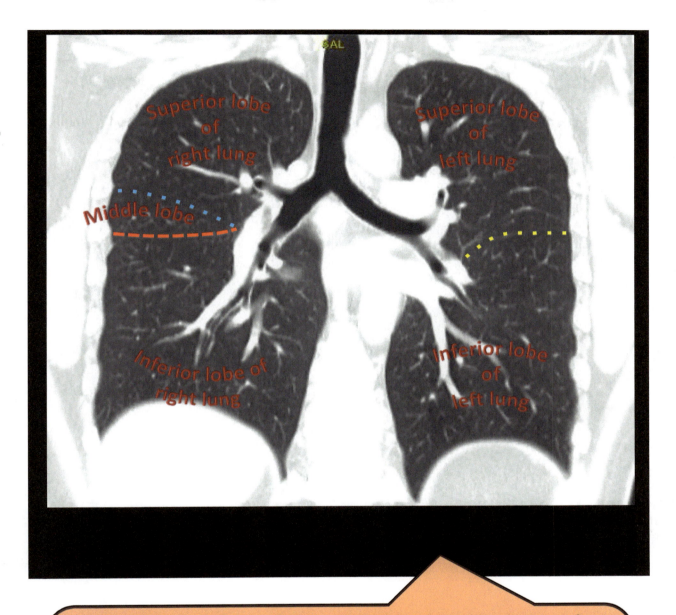

> The right and left major fissures are generally detected as lucent bands, less frequently as *lines*, and rarely as dense bands.
> The minor fissure is more frequently identified as lucent, triangular area with apex at minor fissure. Rarely, it can be observed as a round or oval lucent area [2].

 OBLIQUE FISSURE OF RIGHT LUNG
OBLIQUE FISSURE OF LEFT LUNG
HORIZONTAL FISSURE OF RIGHT LUNG

See Ref [2]

Axial (bronchopulmonary segments)

Thorax Chapter | 1   11

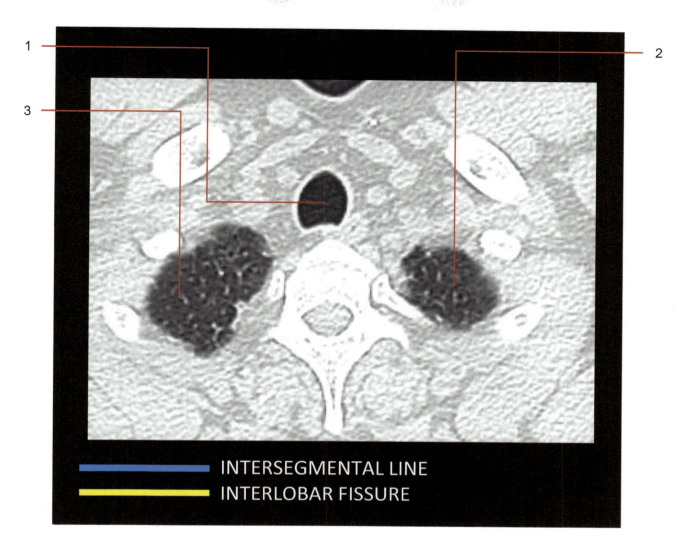

1 Trachea (thoracic part)
2 Apicoposterior segment of left lung
   [S1+2, superior lobe]
3 Apical segment of right lung [S1, superior lobe]

CT LUNG window width (WW) 1500
CT LUNG window level (WL) - 600

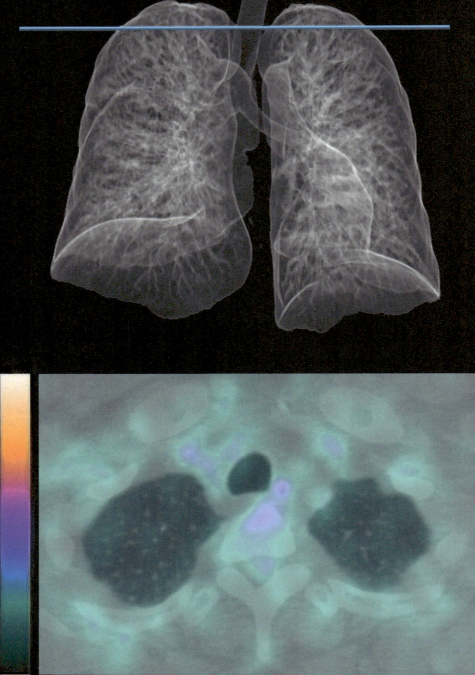

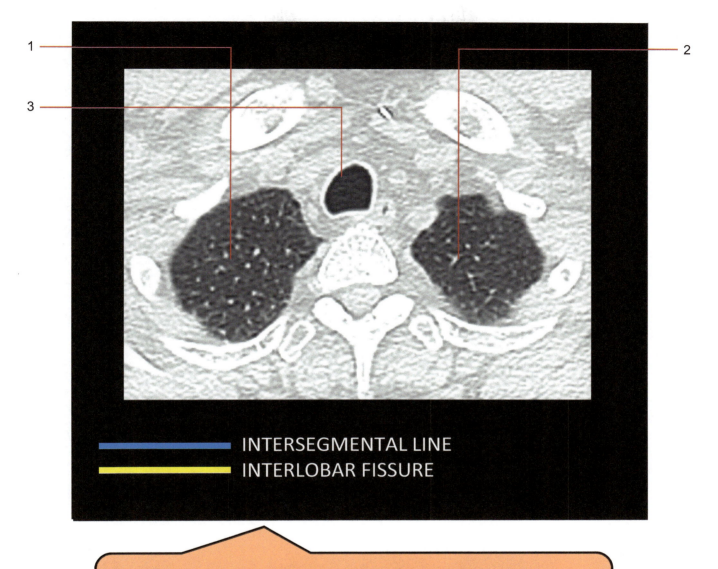

Identification of arteries and their orientation to the plane of the CT section may allow gross identification and localization of bronchopulmonary segments [3].

1 Apical segment of right lung [S1, superior lobe]
2 Apicoposterior segment of left lung [S1+2, superior lobe]
3 Trachea (thoracic part)

See Ref [3]

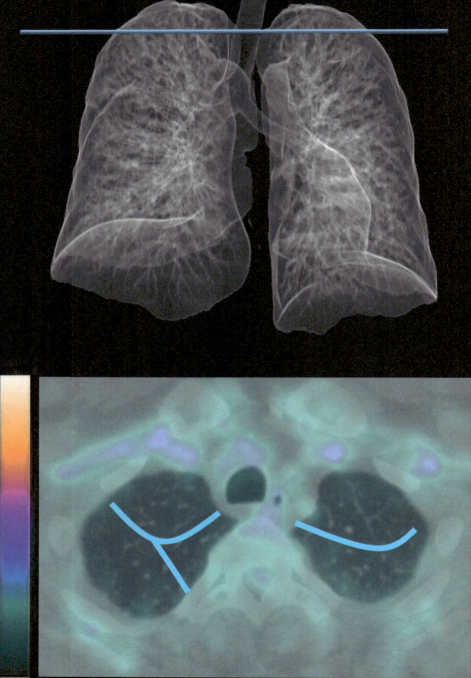

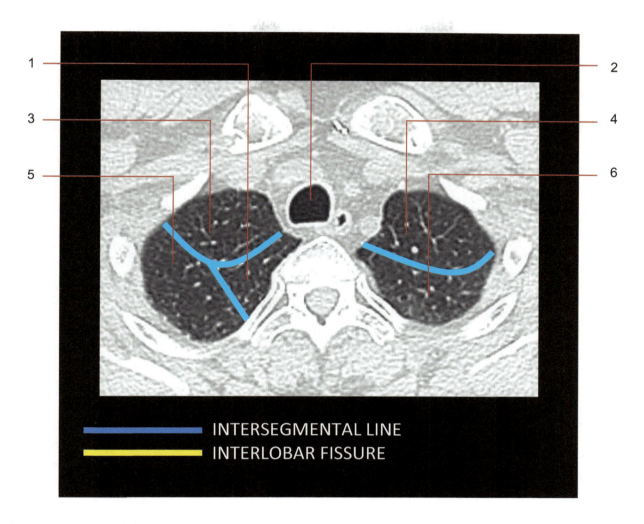

1 Apical segment of right lung [S1, superior lobe]
2 Trachea (thoracic part)
3 Anterior segment of right lung [S3, superior lobe]
4 Anterior segment of left lung [S3, superior lobe]
5 Posterior segment of right lung [S2, superior lobe]
6 Apicoposterior segment of left lung [S1+2, superior lobe]

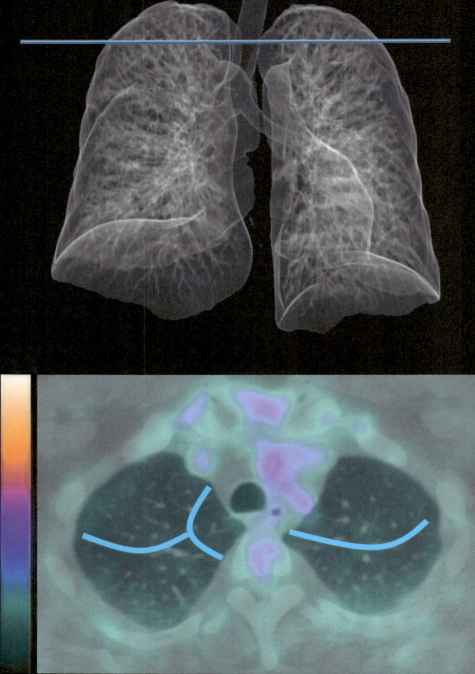

Thorax **Chapter** | 1 17

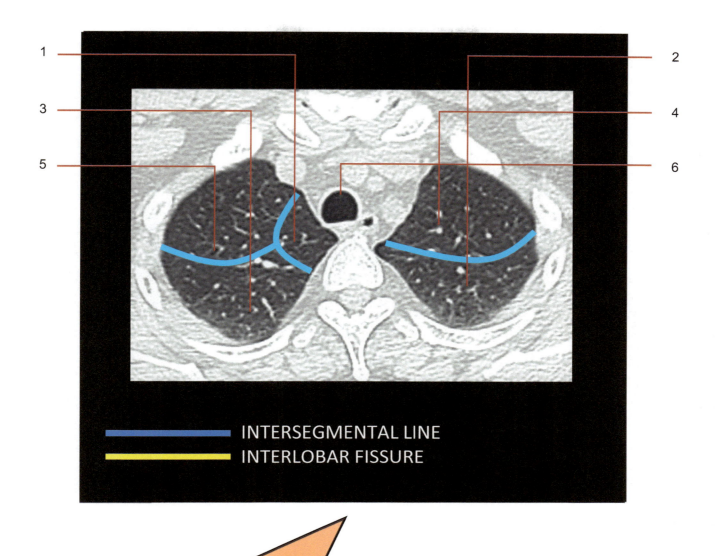

Visual anatomical analysis is at the basis of CT assessment of normal lung parenchima and pathologic findings [4].

1 Apical segment of right lung [S1, superior lobe]
2 Apicoposterior segment of left lung [S1+2, superior lobe]
3 Posterior segment of right lung [S2, superior lobe]
4 Anterior segment of left lung [S3, superior lobe]
5 Anterior segment of right lung [S3, superior lobe]
6 Trachea (thoracic part)

See Ref [4]

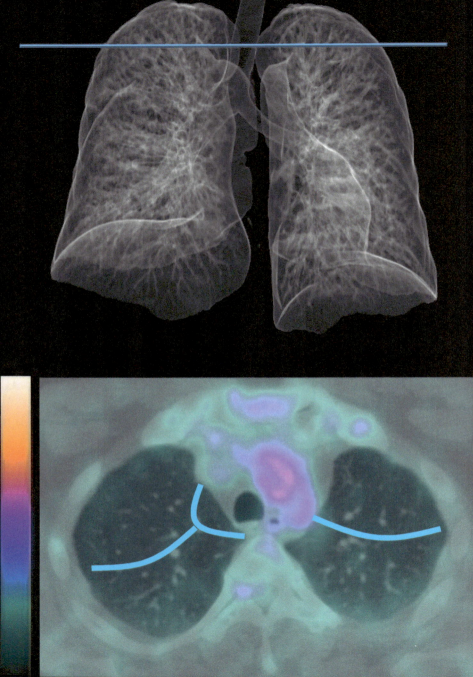

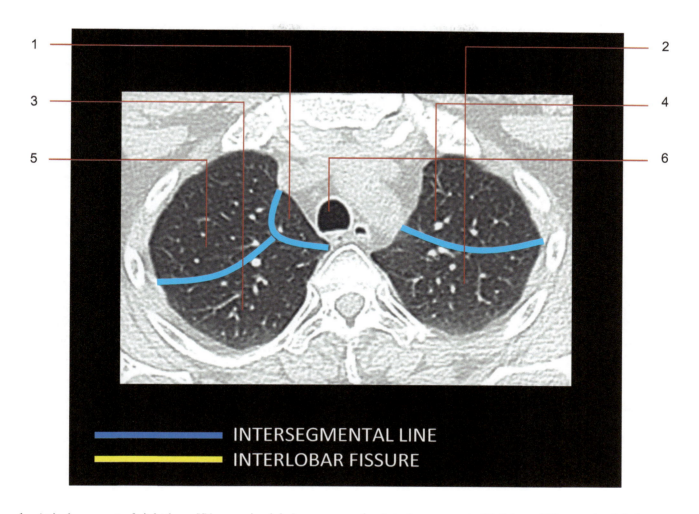

1 Apical segment of right lung [S1, superior lobe]
2 Apicoposterior segment of left lung
  [S1+2, superior lobe]
3 Posterior segment of right lung [S2, superior lobe]
4 Anterior segment of left lung [S3, superior lobe]
5 Anterior segment of right lung [S3, superior lobe]
6 Trachea (thoracic part)

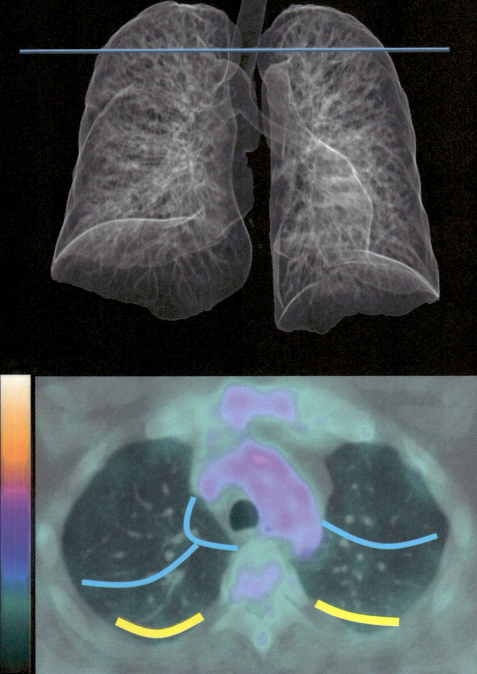

Thorax Chapter | 1   21

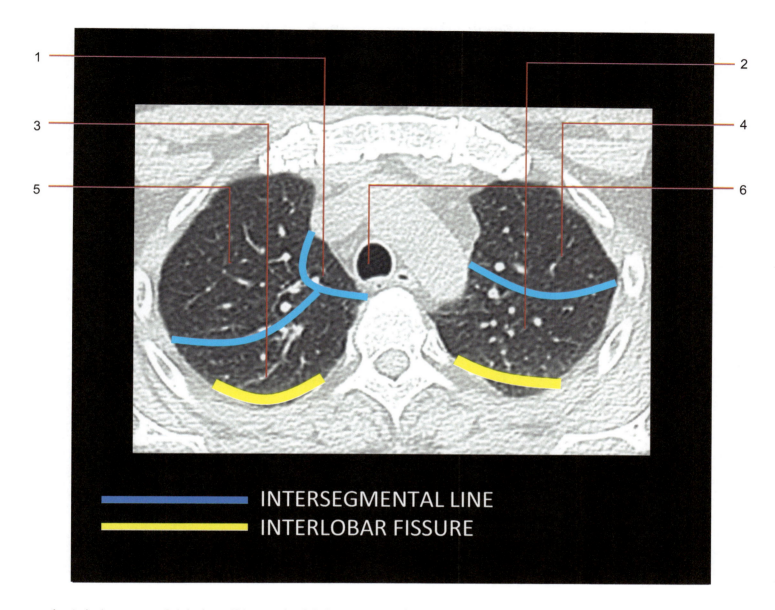

1 Apical segment of right lung [S1, superior lobe]
2 Apicoposterior segment of left lung
   [S1+2, superior lobe]
3 Posterior segment of right lung [S2, superior lobe]
4 Anterior segment of left lung [S3, superior lobe]
5 Anterior segment of right lung [S3, superior lobe]
6 Trachea (thoracic part)

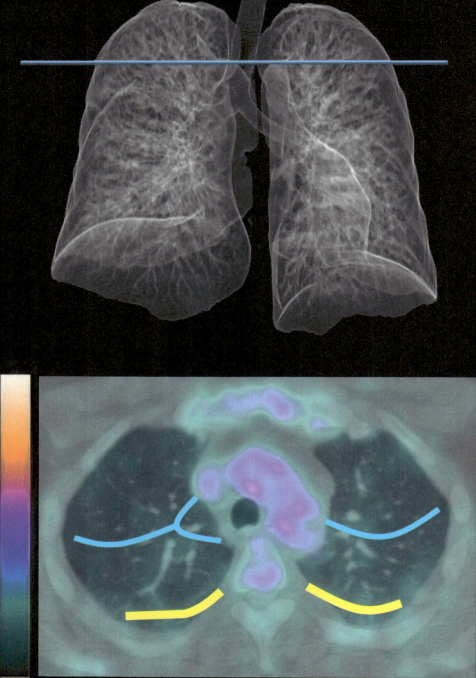

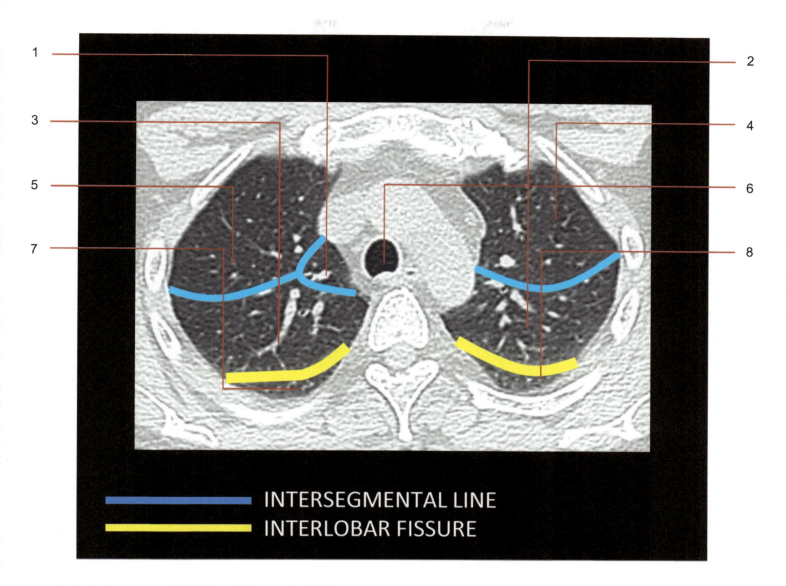

1 Apical segment of right lung [S1, superior lobe]
2 Apicoposterior segment of left lung [S1+2, superior lobe]
3 Posterior segment of right lung [S2, superior lobe]
4 Anterior segment of left lung [S3, superior lobe]
5 Anterior segment of right lung [S3, superior lobe]
6 Trachea (thoracic part)
7 Superior segment of right lung [S6, inferior lobe]
8 Superior segment of left lung [S6, inferior lobe]

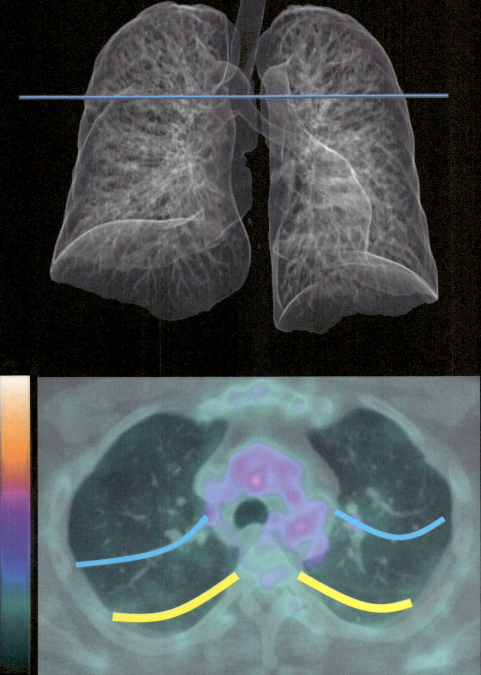

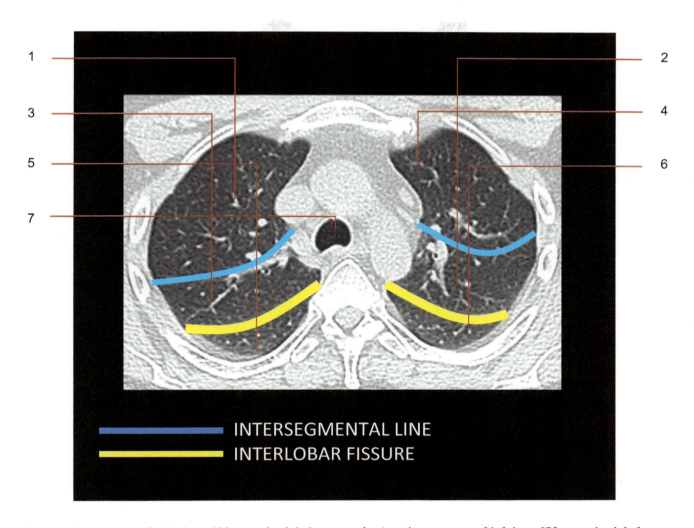

1 Anterior segment of right lung [S3, superior lobe]
2 Apicoposterior segment of left lung [S1+2, superior lobe]
3 Posterior segment of right lung [S2, superior lobe]
4 Anterior segment of left lung [S3, superior lobe]
5 Superior segment of right lung [S6, inferior lobe]
6 Superior segment of left lung [S6, inferior lobe]
7 Trachea (thoracic part)

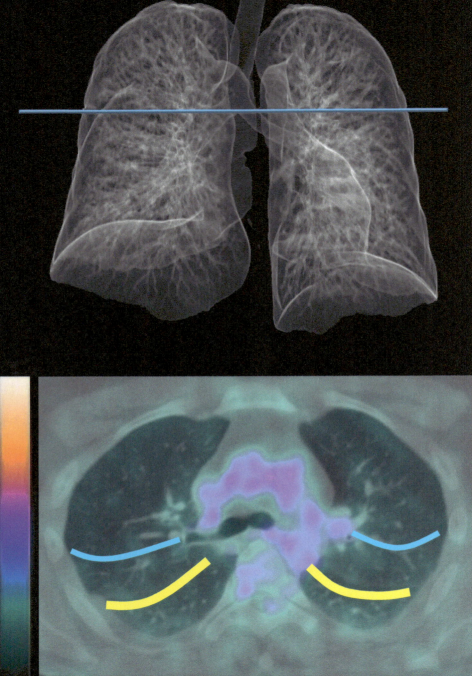

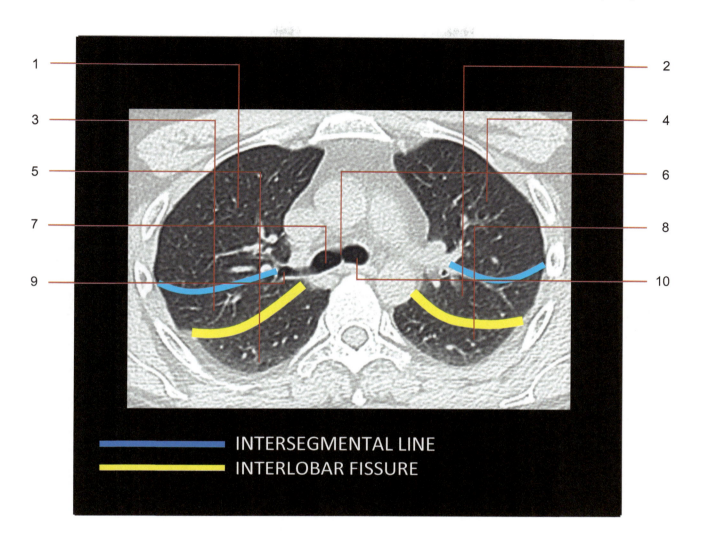

1 Anterior segment of right lung [S3, superior lobe]
2 Apicoposterior segment of left lung [S1+2, superior lobe]
3 Posterior segment of right lung [S2, superior lobe]
4 Anterior segment of left lung [S3, superior lobe]
5 Superior segment of right lung [S6, inferior lobe]
6 Carina of trachea
7 Right main bronchus
8 Superior segment of left lung [S6, inferior lobe]
9 Right superior lobar bronchus
10 Left main bronchus

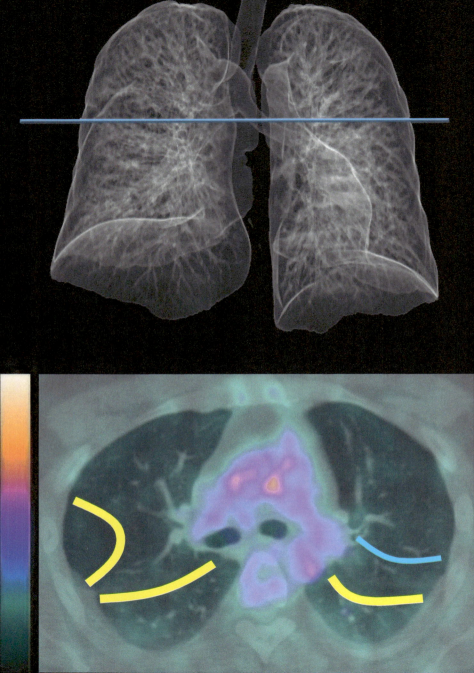

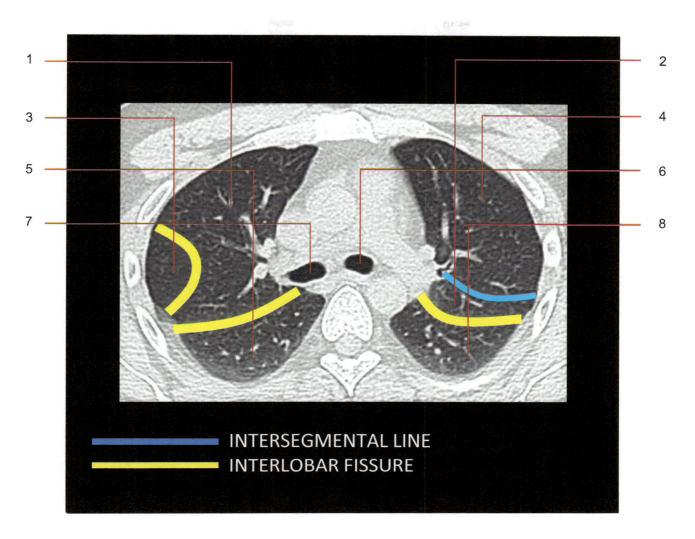

1 Anterior segment of right lung [S3, superior lobe]
2 Apicoposterior segment of left lung [S1+2, superior lobe]
3 Lateral segment of right lung [S4, middle lobe]
4 Anterior segment of left lung [S3, superior lobe]
5 Superior segment of right lung [S6, inferior lobe]
6 Left main bronchus
7 Intermediate bronchus
8 Superior segment of left lung [S6, inferior lobe]

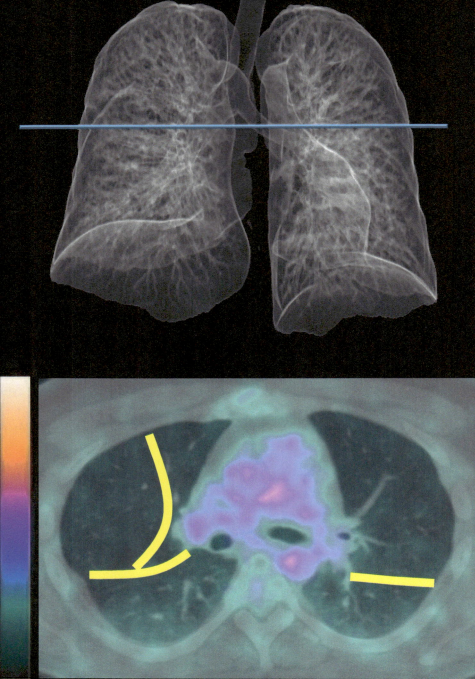

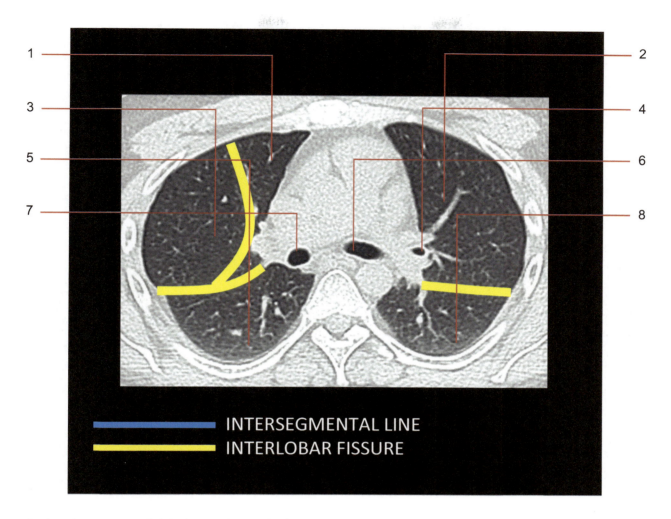

1 Anterior segment of right lung [S3, superior lobe]
2 Anterior segment of left lung [S3, superior lobe]
3 Lateral segment of right lung [S4, middle lobe]
4 Left superior lobar bronchus
5 Superior segment of right lung [S6, inferior lobe]
6 Left main bronchus
7 Intermediate bronchus
8 Superior segment of left lung [S6, inferior lobe]

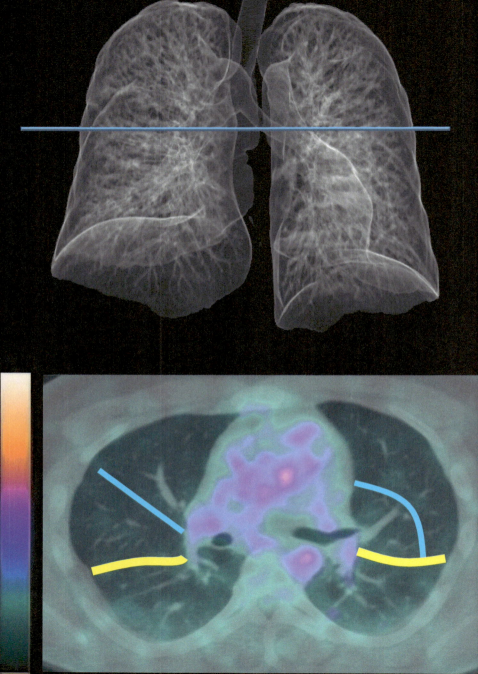

Thorax **Chapter | 1** 33

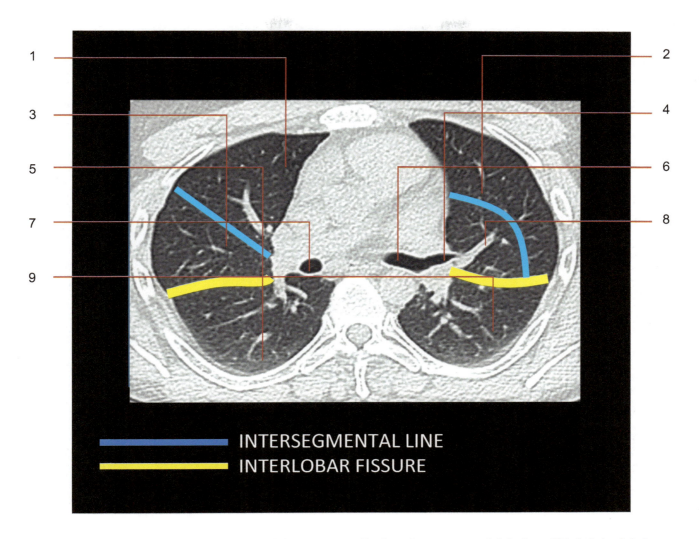

1 Medial segment of right lung [S5, middle lobe]
2 Superior lingular segment of left lung [S4, superior lobe]
3 Lateral segment of right lung [S4, middle lobe]
4 Lingular segmental bronchus of left lung
5 Superior segment of right lung [S6, inferior lobe]
6 Left main bronchus
7 Intermediate bronchus
8 Inferior lingular segment of left lung [S5, superior lobe]
9 Superior segment of left lung [S6, inferior lobe]

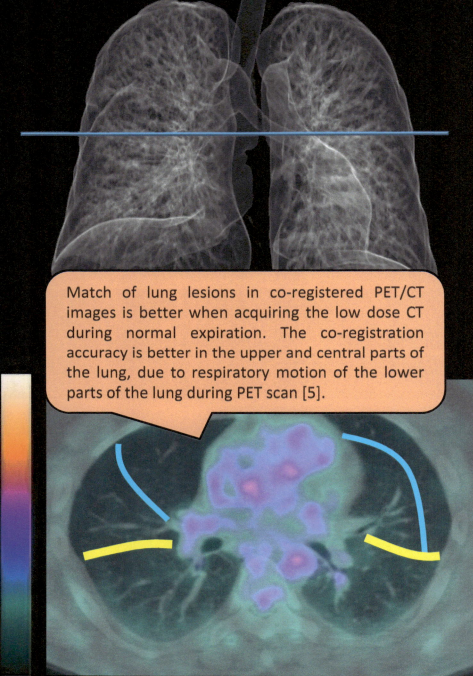

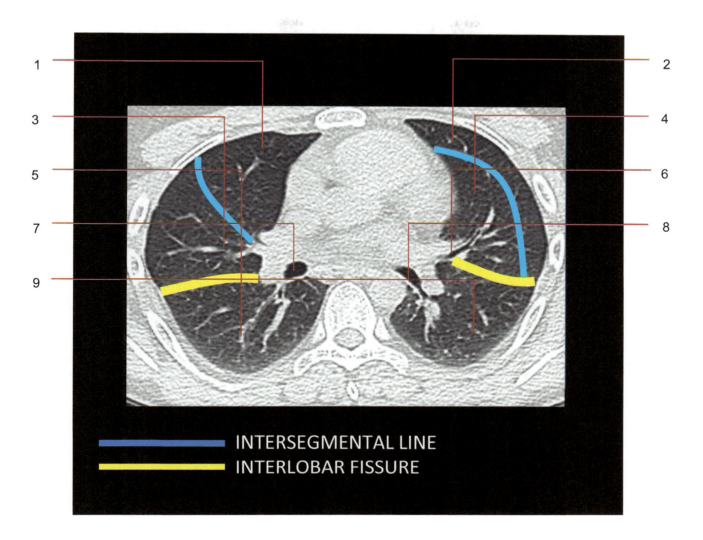

1 Medial segment of right lung [S5, middle lobe]
2 Superior lingular segment [S4, superior lobe]
3 Lateral segment of right lung [S4, middle lobe]
4 Inferior lingular segment of left lung [S5, superior lobe]
5 Superior segment of right lung [S6, inferior lobe]
6 Lingular segmental bronchus of left lung
7 Intermediate bronchus
8 Left inferior lobar bronchus
9 Superior segment of left lung [S6, inferior lobe]

See Ref [5]

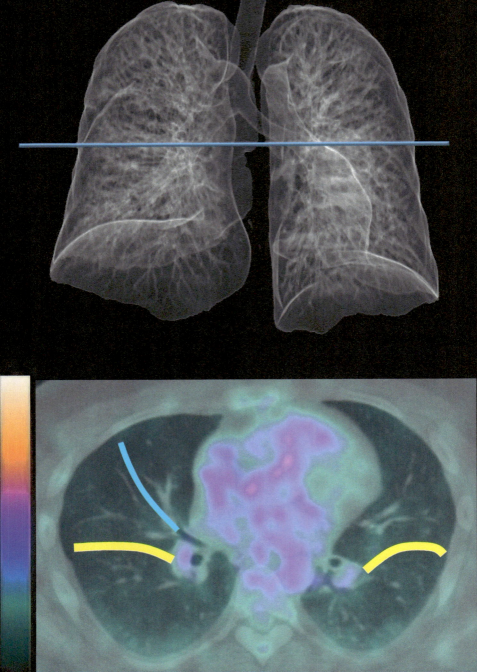

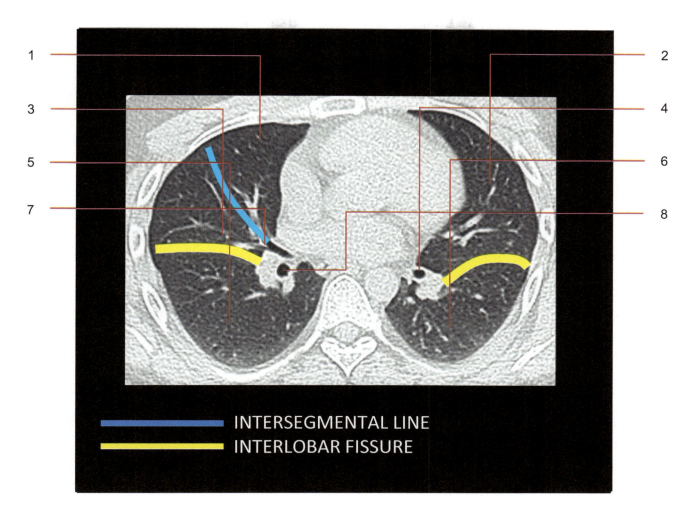

1 Medial segment of right lung [S5, middle lobe]
2 Inferior lingular segment of left lung [S5, superior lobe]
3 Lateral segment of right lung [S4, middle lobe]
4 Left inferior lobar bronchus
5 Superior segment of right lung [S6, inferior lobe]
6 Superior segment of left lung [S6, inferior lobe]
7 Middle lobar bronchus
8 Right inferior lobar bronchus

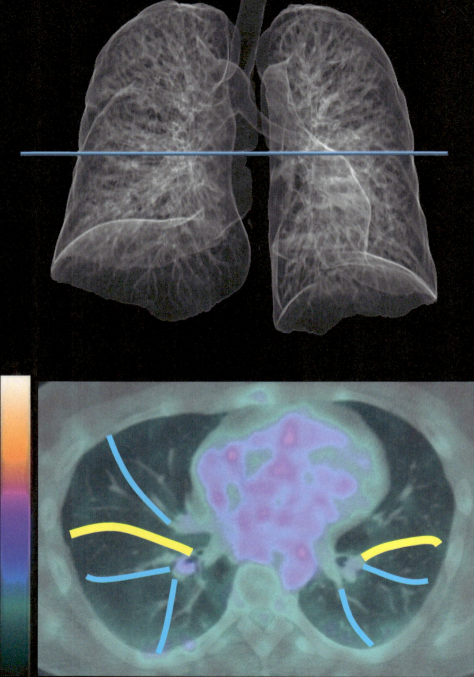

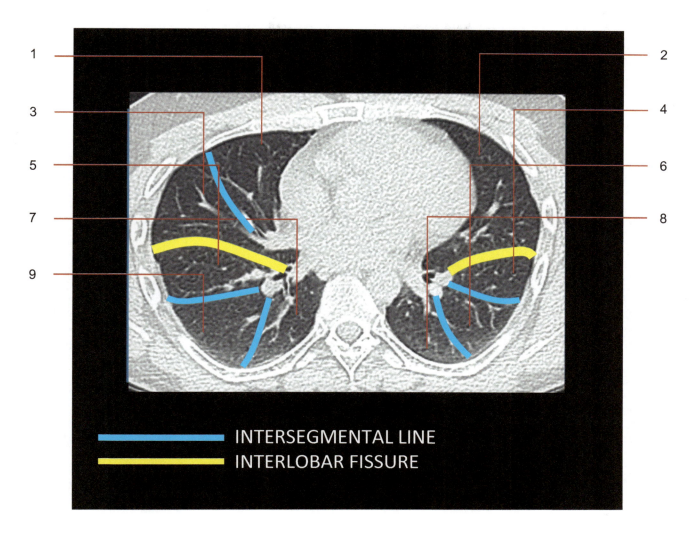

1 Medial segment of right lung [S5, middle lobe]
2 Inferior lingular segment of left lung [S5, superior lobe]
3 Lateral segment of right lung [S4, middle lobe]
4 Anterior basal segment of left lung [S8, inferior lobe]
5 Anterior basal segment of right lung [S8, inferior lobe]
6 Lateral basal segment of left lung [S9, inferior lobe]
7 Posterior basal segment of right lung [S10, inferior lobe]
8 Posterior basal segment of left lung [S10, inferior lobe]

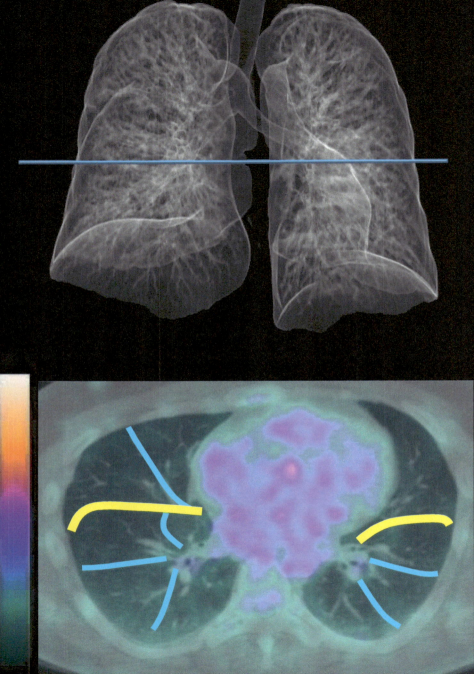

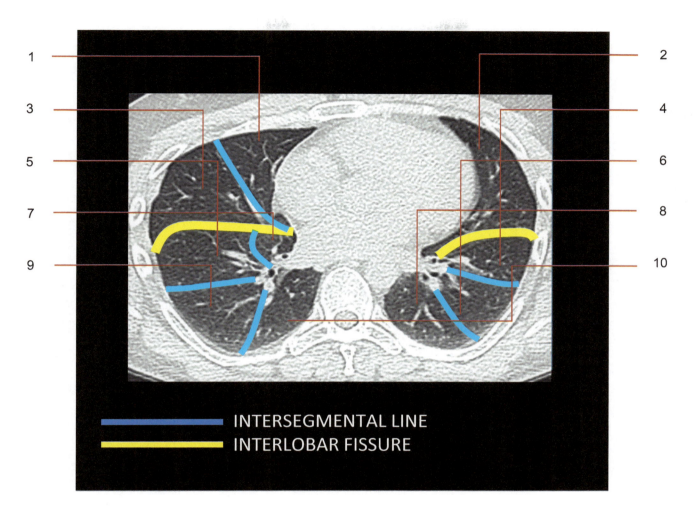

1 Medial segment of right lung [S5, middle lobe]
2 Inferior lingular segment of left lung [S5, superior lobe]
3 Lateral segment of right lung [S4, middle lobe]
4 Anterior basal segment of left lung [S8, inferior lobe]
5 Anterior basal segment of right lung [S8, inferior lobe]
6 Lateral basal segment of left lung [S9, inferior lobe]
7 Medial basal segment of right lung [S7, inferior lobe]
8 Posterior basal segment of left lung [S10, inferior lobe]
9 Lateral basal segment of right lung [S9, inferior lobe]
10 Posterior basal segment of right lung [S10, inferior lobe]

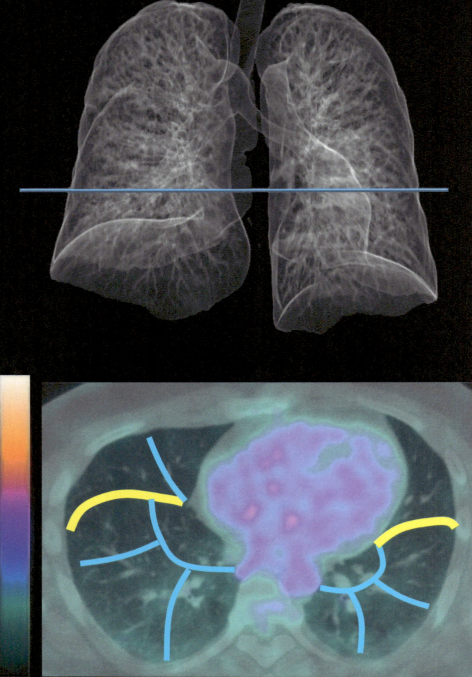

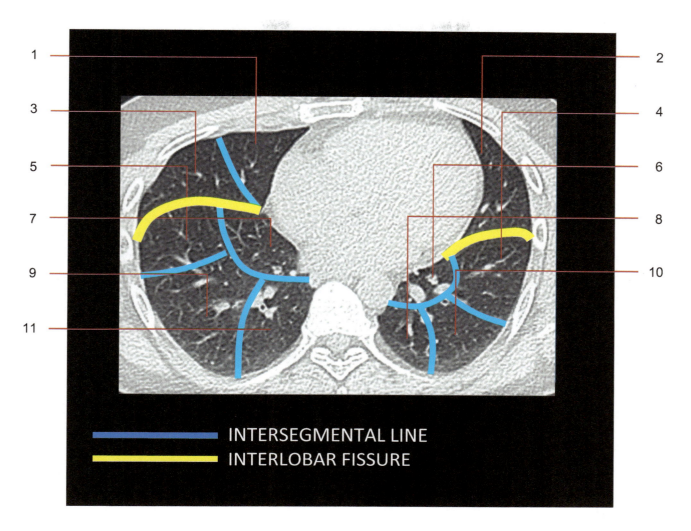

1 Medial segment of right lung [S5, middle lobe]
2 Inferior lingular segment of left lung [S5, superior lobe]
3 Lateral segment of right lung [S4, middle lobe]
4 Anterior basal segment of left lung [S8, inferior lobe]
5 Anterior basal segment of right lung [S8, inferior lobe]
6 Medial basal segment of left lung [S7, inferior lobe]
7 Medial basal segment of right lung [S7, inferior lobe]
8 Posterior basal segment of left lung [S10, inferior lobe]
9 Lateral basal segment of right lung [S9, inferior lobe]
10 Lateral basal segment of left lung [S9, inferior lobe]
11 Posterior basal segment of right lung [S10, inferior lobe]

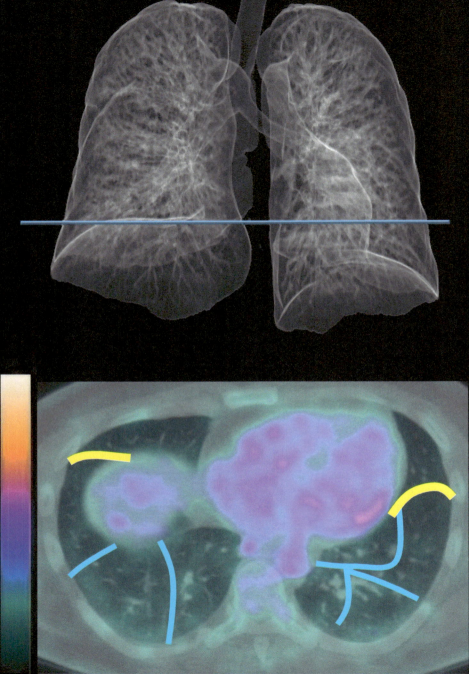

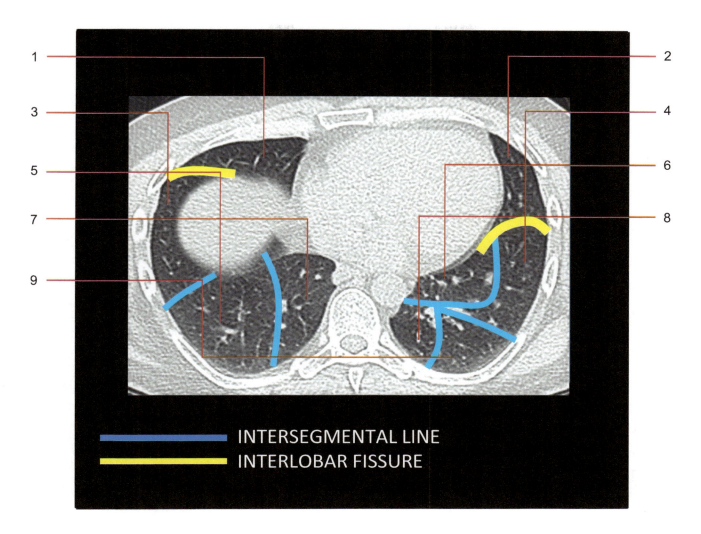

1 Medial segment of right lung [S5, middle lobe]
2 Inferior lingular segment of left lung [S5, superior lobe]
3 Anterior basal segment of right lung [S8, inferior lobe]
4 Anterior basal segment of left lung [S8, inferior lobe]
5 Lateral basal segment of right lung [S9, inferior lobe]
6 Medial basal segment of left lung [S7, inferior lobe]
7 Posterior basal segment of right lung [S10, inferior lobe]
8 Posterior basal segment of left lung [S10, inferior lobe]
9 Lateral basal segment of left lung [S9, inferior lobe]

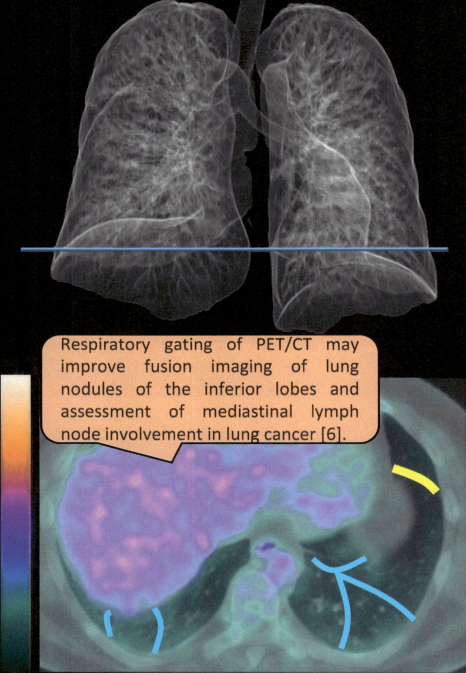

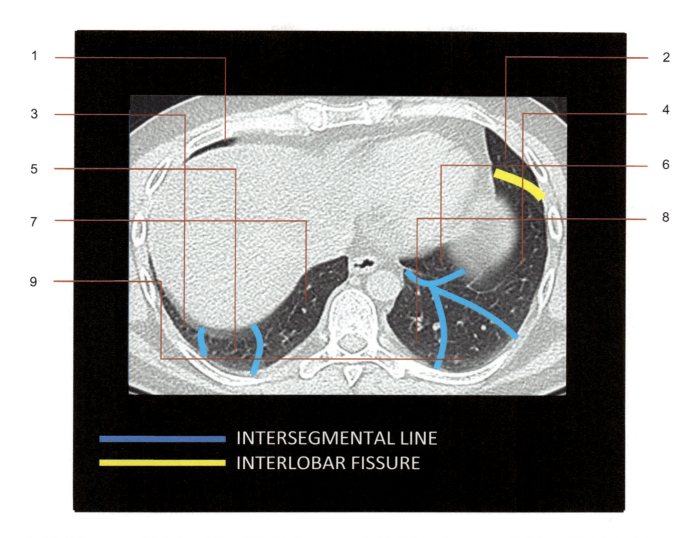

1. Medial segment of right lung [S5, middle lobe]
2. Inferior lingular segment of left lung [S5, superior lobe]
3. Anterior basal segment of right lung [S8, inferior lobe]
4. Anterior basal segment of left lung [S8, inferior lobe]
5. Lateral basal segment of right lung [S9, inferior lobe]
6. Medial basal segment of left lung [S7, inferior lobe]
7. Posterior basal segment of right lung [S10, inferior lobe]
8. Posterior basal segment of left lung [S10, inferior lobe]
9. Lateral basal segment of left lung [S9, inferior lobe]

See Ref [6]

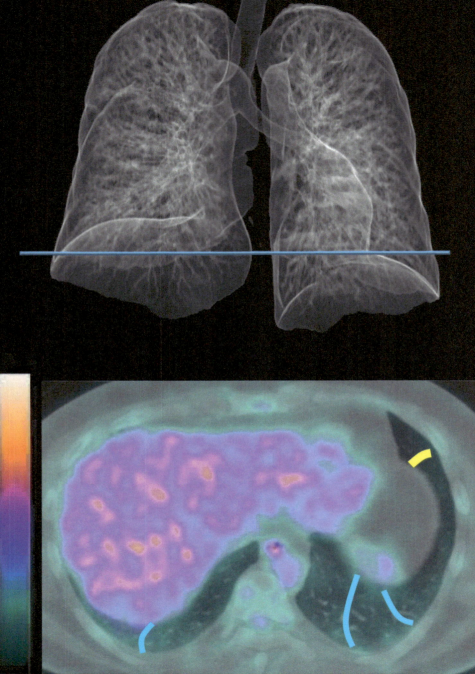

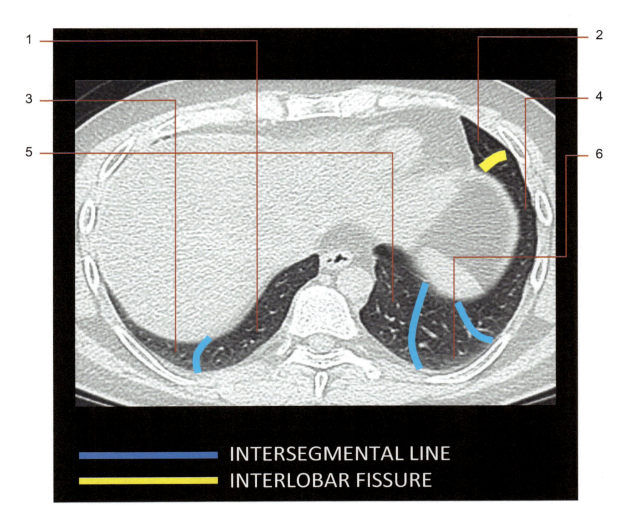

INTERSEGMENTAL LINE
INTERLOBAR FISSURE

1 Posterior basal segment of right lung [S10, inferior lobe]
2 Inferior lingular segment of left lung [S5, superior lobe]
3 Lateral basal segment of right lung [S9, inferior lobe]
4 Anterior basal segment of left lung [S8, inferior lobe]
5 Posterior basal segment of left lung [S10, inferior lobe]
6 Lateral basal segment of left lung [S9, inferior lobe]

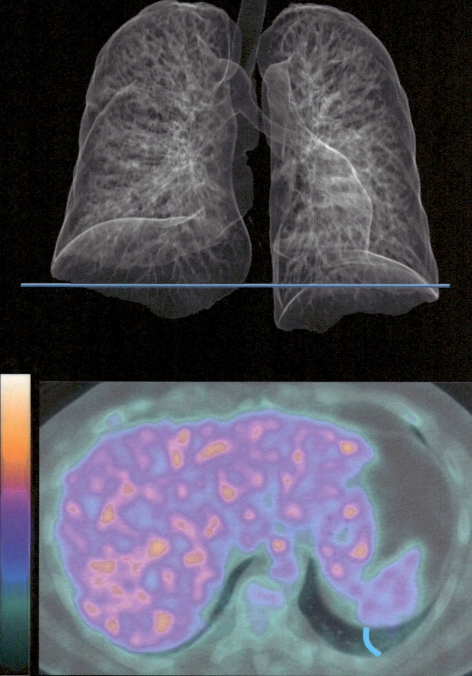

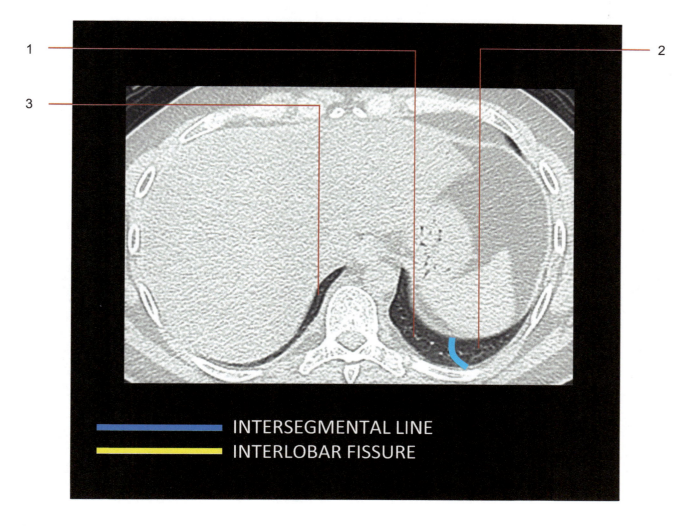

1 Posterior basal segment of left lung [S10, inferior lobe]
2 Lateral basal segment of left lung [S9, inferior lobe]
3 Posterior basal segment of right lung [S10, inferior lobe]

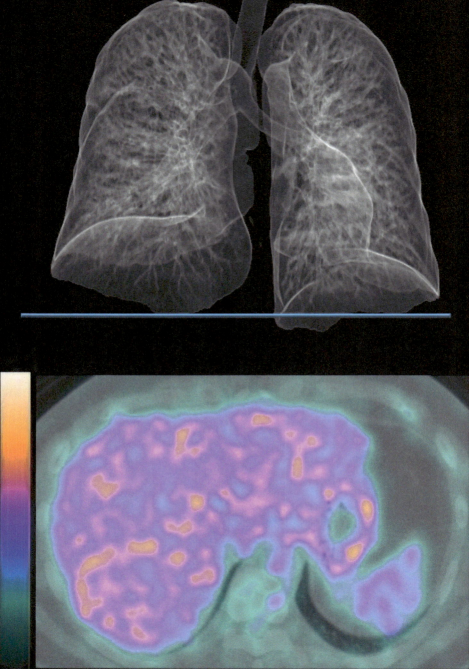

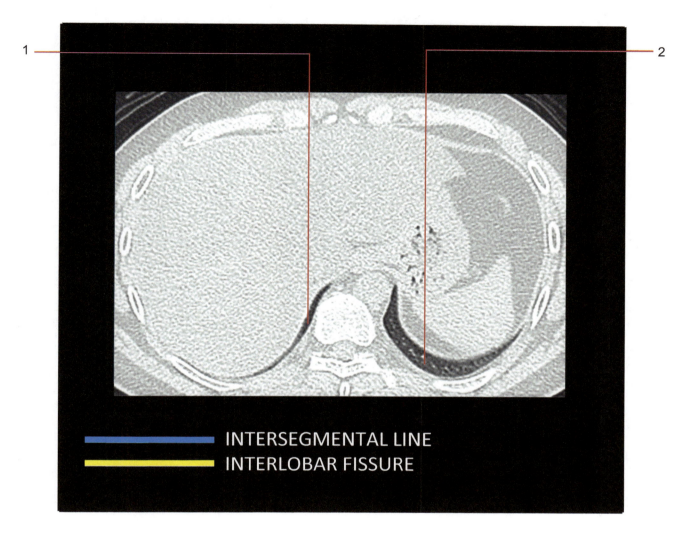

1 Posterior basal segment of right lung [S10, inferior lobe]   **2** Posterior basal segment of left lung [S10, inferior lobe]

1.1.3 Sagittal (bronchopulmonary segments)

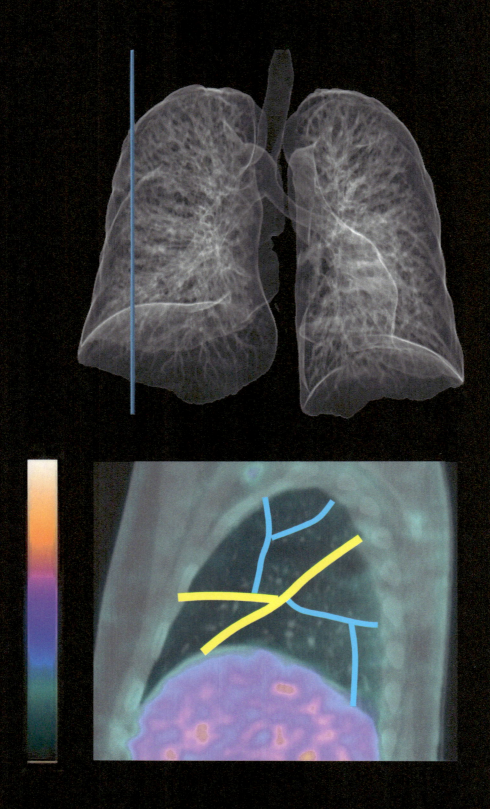

Thorax Chapter | 1   55

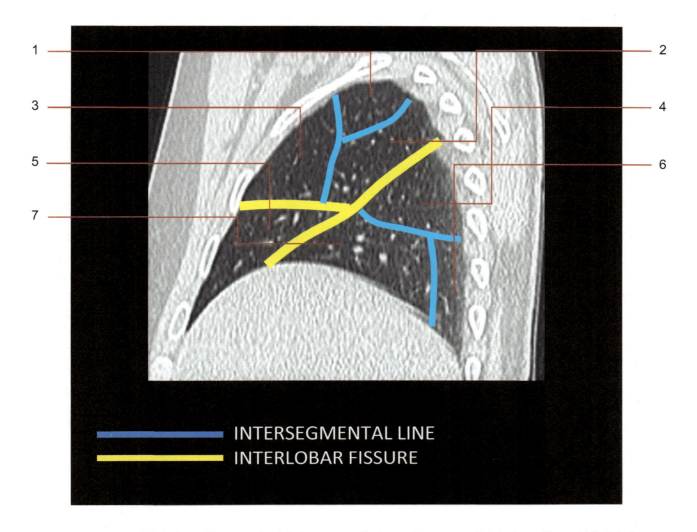

1 Apical segment of right lung [S1, superior lobe]
2 Posterior segment of right lung [S2, superior lobe]
3 Anterior segment of right lung [S3, superior lobe]
4 Superior segment of right lung [S6, inferior lobe]
5 Lateral segment of right lung [S4, middle lobe]
6 Lateral basal segment of right lung [S9, inferior lobe]
7 Anterior basal segment of right lung [S8, inferior lobe]

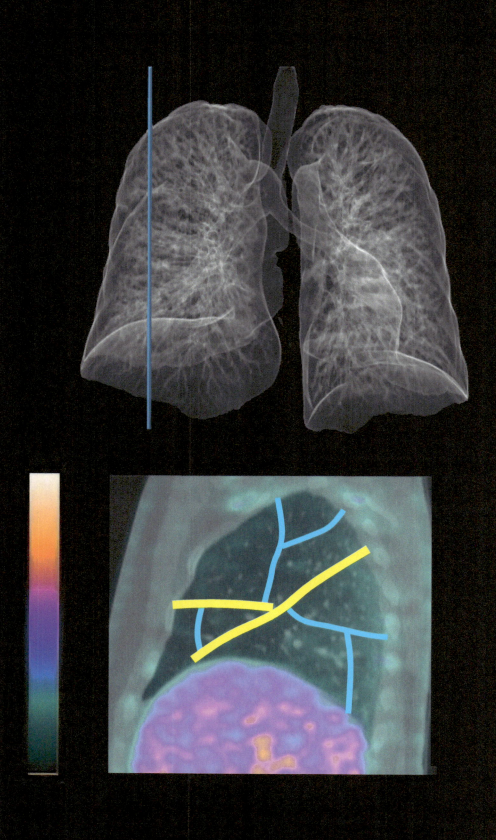

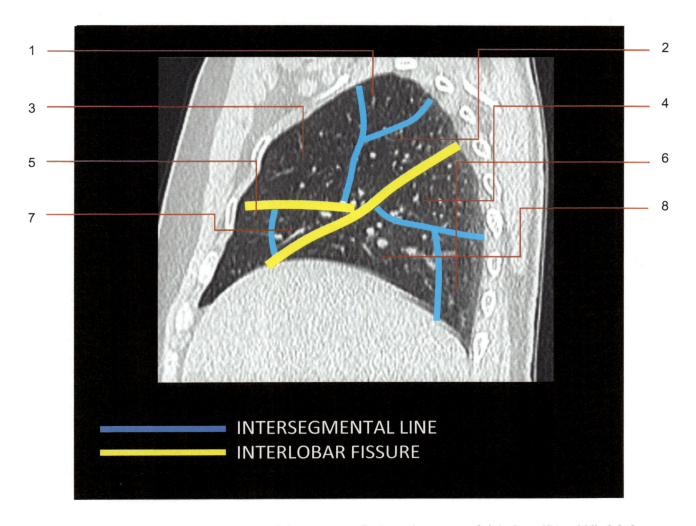

1 Apical segment of right lung [S1, superior lobe]
2 Posterior segment of right lung [S2, superior lobe]
3 Anterior segment of right lung [S3, superior lobe]
4 Superior segment of right lung [S6, inferior lobe]
5 Lateral segment of right lung [S4, middle lobe]
6 Lateral basal segment of right lung [S9, inferior lobe]
7 Medial segment of right lung [S5, middle lobe]
8 Anterior basal segment of right lung [S8, inferior lobe]

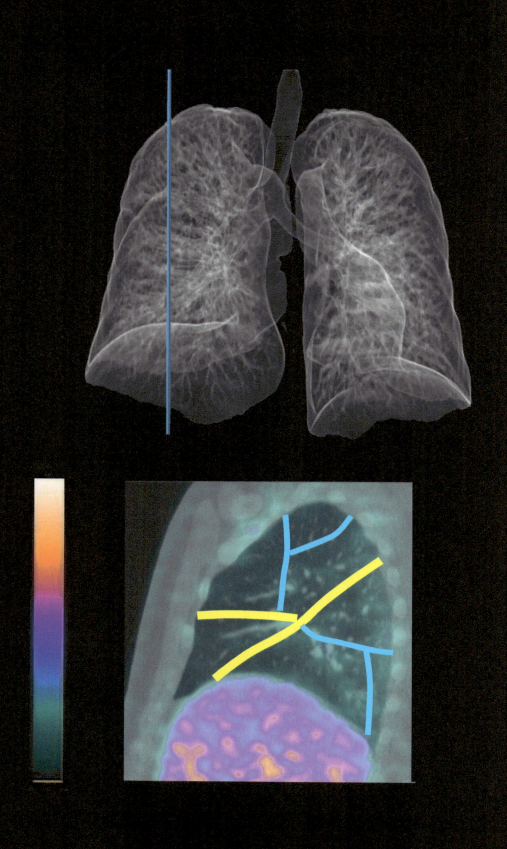

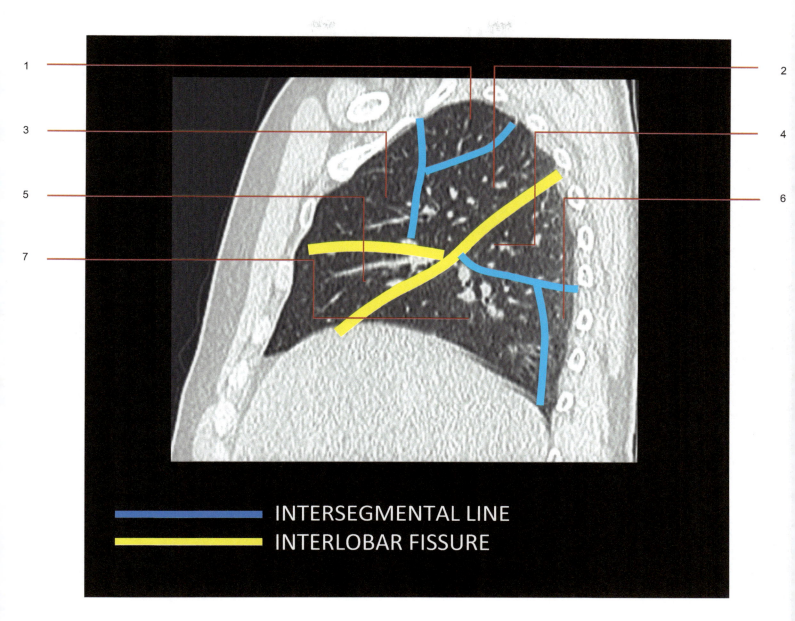

1 Apical segment of right lung [S1, superior lobe]
2 Posterior segment of right lung [S2, superior lobe]
3 Anterior segment of right lung [S3, superior lobe]
4 Superior segment of right lung [S6, inferior lobe]
5 Medial segment of right lung [S5, middle lobe]
6 Lateral basal segment of right lung [S9, inferior lobe]
7 Anterior basal segment of right lung [S8, inferior lobe]

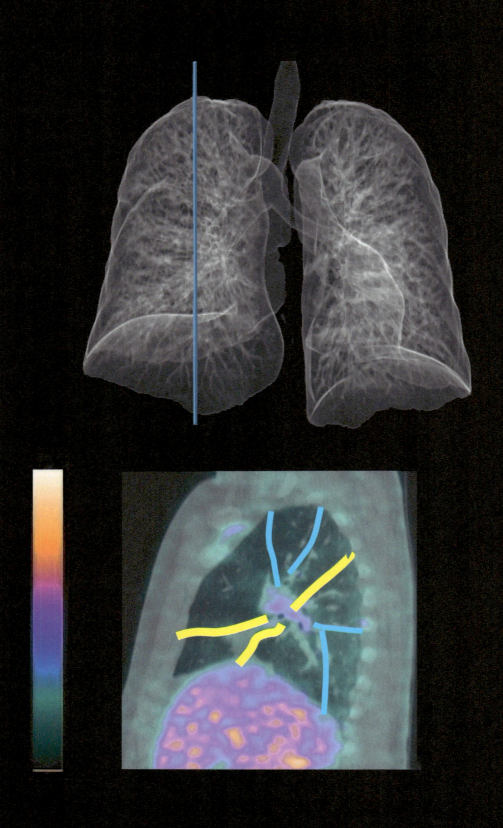

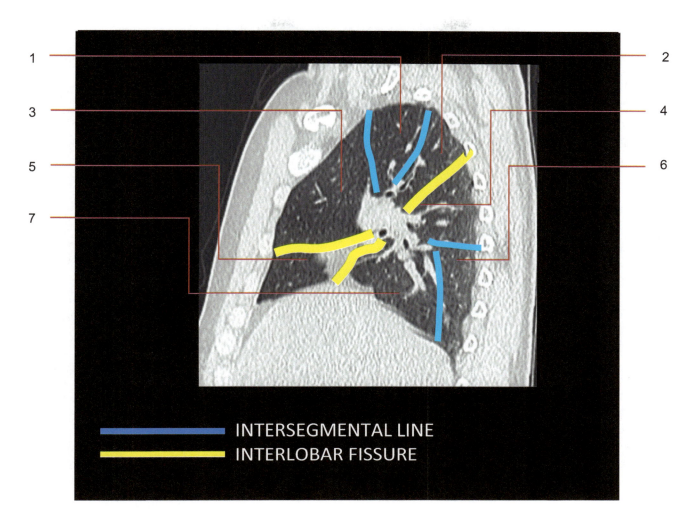

1 Apical segment of right lung [S1, superior lobe]
2 Posterior segment of right lung [S2, superior lobe]
3 Anterior segment of right lung [S3, superior lobe]
4 Superior segment of right lung [S6, inferior lobe]
5 Medial segment of right lung [S5, middle lobe]
6 Posterior basal segment of right lung [S10, inferior lobe]
7 Anterior basal segment of right lung [S8, inferior lobe]

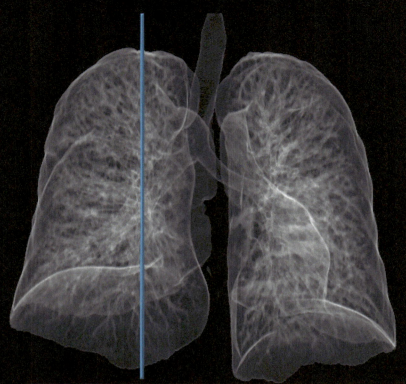

Respiratory motion artifacts may affect whole body 18F-FDG PET/CT, especially for lesions between liver and lungs [7].

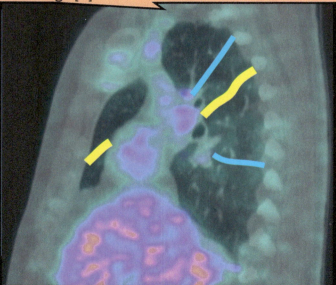

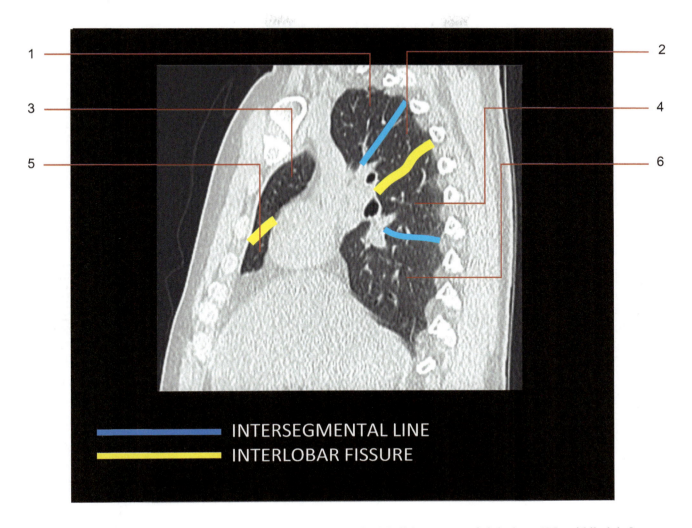

1 Apical segment of right lung [S1, superior lobe]
2 Posterior segment of right lung [S2, superior lobe]
3 Anterior segment of right lung [S3, superior lobe]
4 Superior segment of right lung [S6, inferior lobe]
5 Medial segment of right lung [S5, middle lobe]
6 Posterior basal segment of right lung [S10, inferior lobe]

See Ref [7]

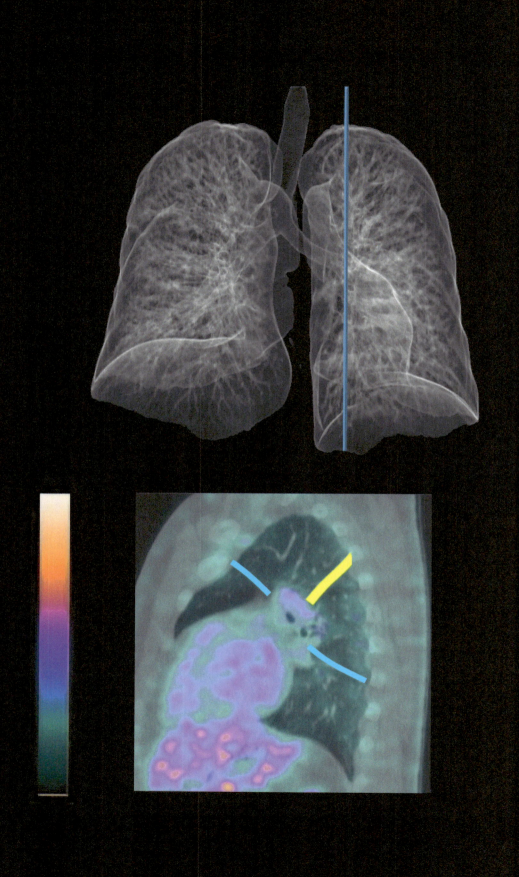

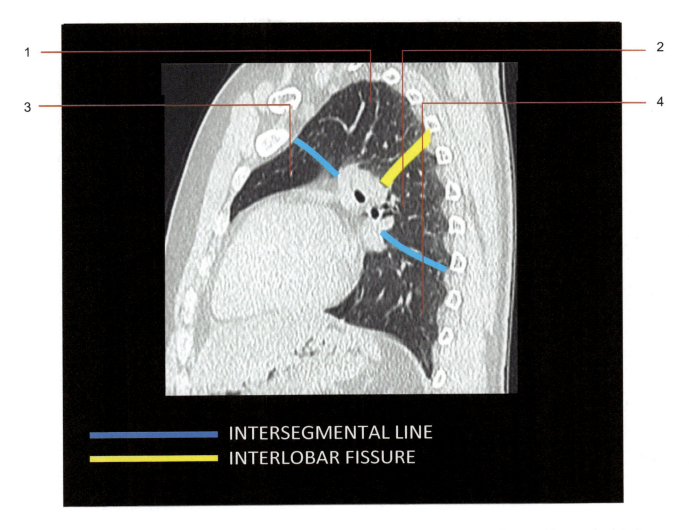

1 Apicoposterior segment of left lung [S1+2, superior lobe]
2 Superior segment of left lung [S6, inferior lobe]
3 Anterior segment of left lung [S3, superior lobe]
4 Posterior basal segment of left lung [S10, inferior lobe]

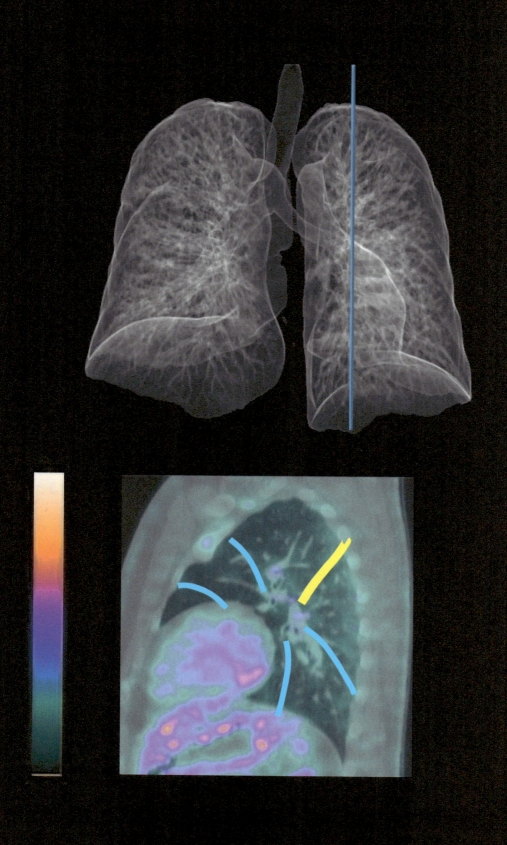

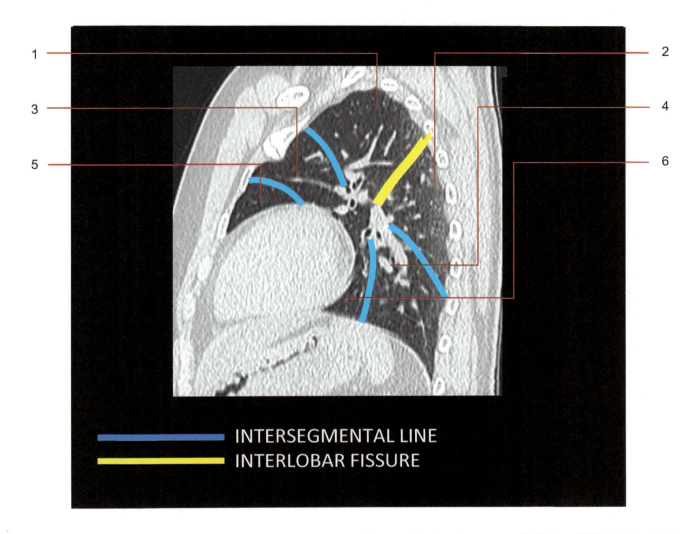

1 Apicoposterior segment of left lung [S1+2, superior lobe]
2 Superior segment of left lung [S6, inferior lobe]
3 Anterior segment of left lung [S3, superior lobe]
4 Posterior basal segment of left lung [S10, inferior lobe]
5 Superior lingular segment of left lung [S4, superior lobe]
6 Medial basal segment of left lung [S7, inferior lobe]

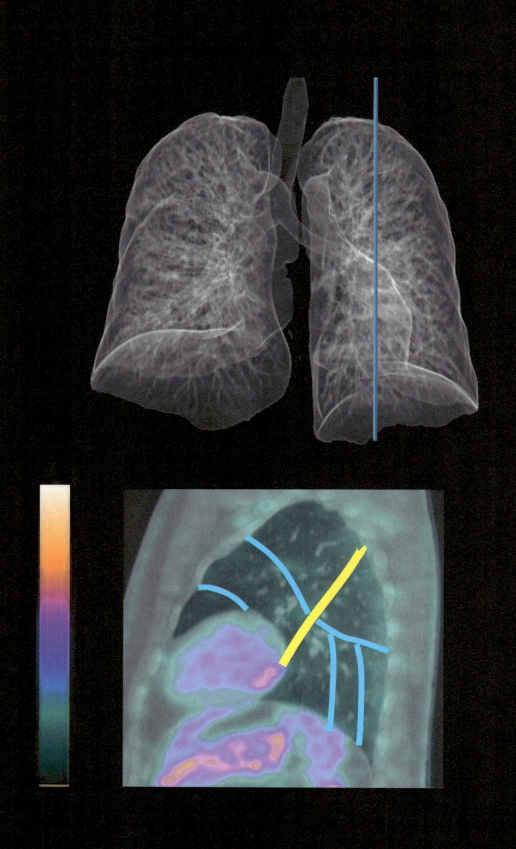

Thorax **Chapter** | 1   69

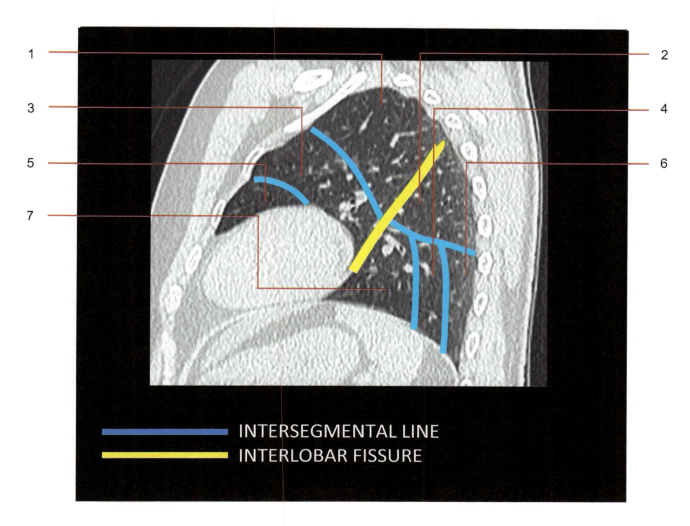

INTERSEGMENTAL LINE
INTERLOBAR FISSURE

1   Apicoposterior segment of left lung [S1+2, superior lobe]
2   Superior segment of left lung [S6, inferior lobe]
3   Anterior segment of left lung [S3, superior lobe]
4   Lateral basal segment of left lung [S9, inferior lobe]
5   Superior lingular segment of left lung [S4, superior lobe]
6   Posterior basal segment of left lung [S10, inferior lobe]
7   Anterior basal segment of left lung [S8, inferior lobe]

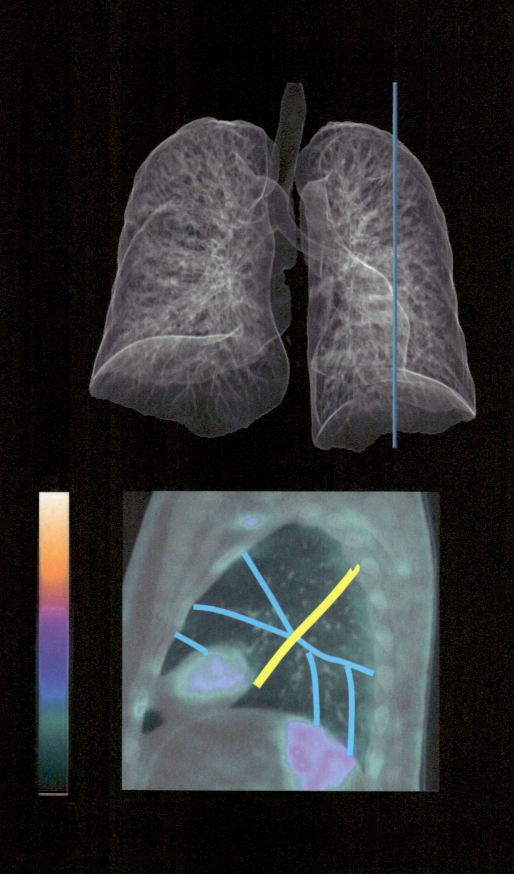

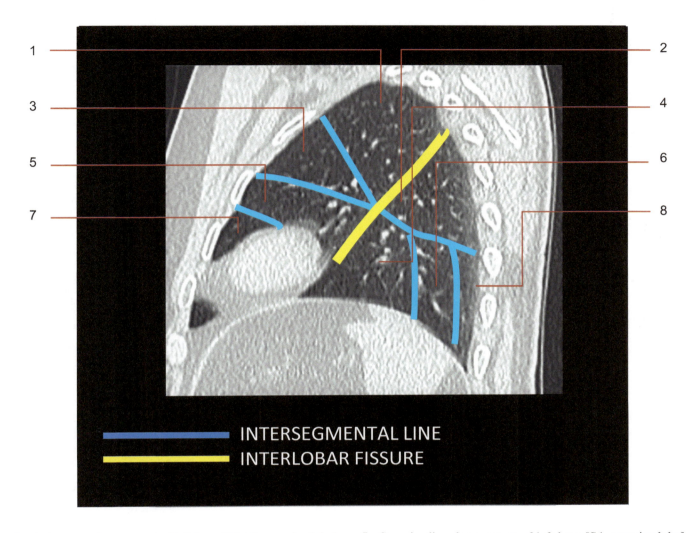

1 Apicoposterior segment of left lung [S1+2, superior lobe]
2 Superior segment of left lung [S6, inferior lobe]
3 Anterior segment of left lung [S3, superior lobe]
4 Anterior basal segment of left lung [S8, inferior lobe]
5 Superior lingular segment of left lung [S4, superior lobe]
6 Lateral basal segment of left lung [S9, inferior lobe]
7 Inferior lingular segment of left lung [S5, superior lobe]
8 Posterior basal segment of left lung [S10, inferior lobe]

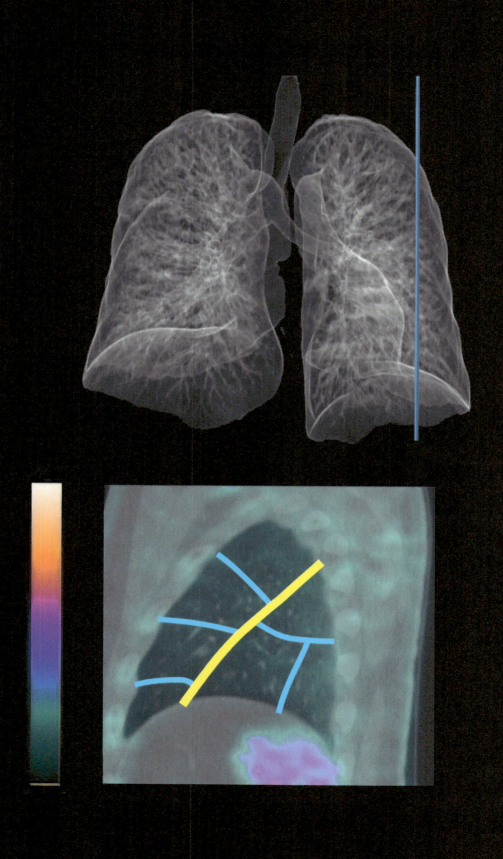

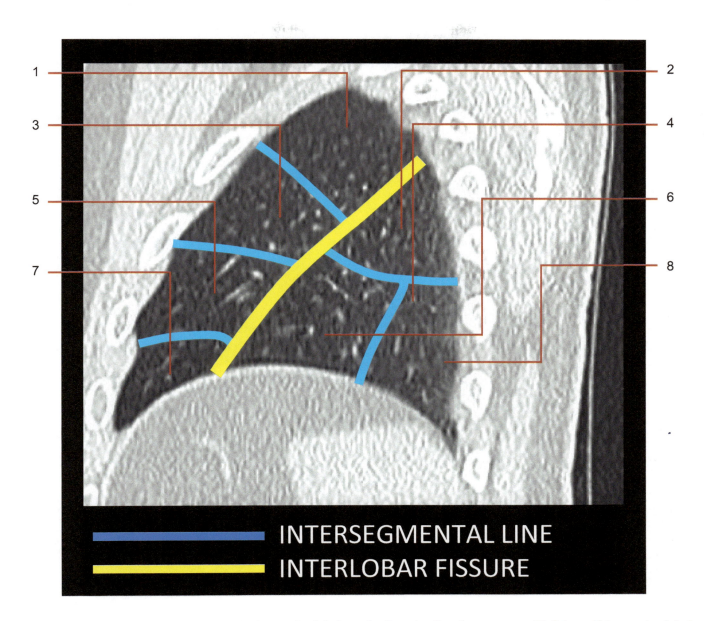

1 Apicoposterior segment of left lung [S1+2, superior lobe]
2 Superior segment of left lung [S6, inferior lobe]
3 Anterior segment of left lung [S3, superior lobe]
4 Posterior basal segment of left lung [S10, inferior lobe]
5 Superior lingular segment of left lung [S4, superior lobe]
6 Anterior basal segment of left lung [S8, inferior lobe]
7 Inferior lingular segment of left lung [S5, superior lobe]
8 Lateral basal segment of left lung [S9, inferior lobe]

Coronal (bronchopulmonary segments)

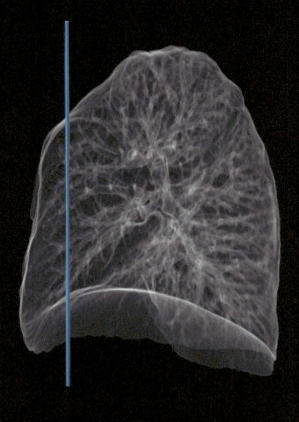
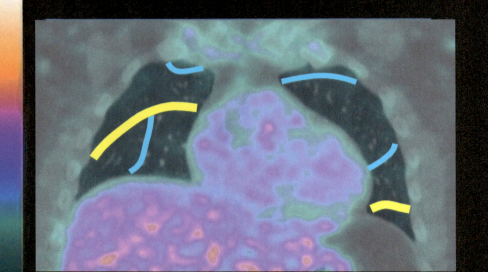

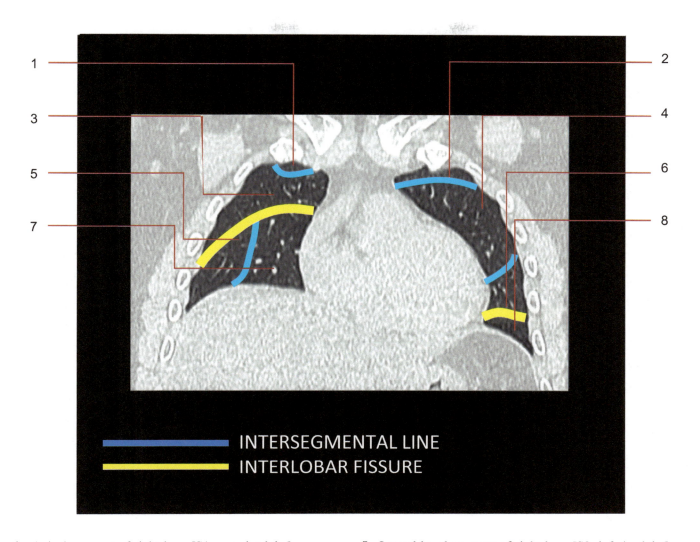

1 Apical segment of right lung [S1, superior lobe]
2 Apicoposterior segment of left lung [S1+2, superior lobe]
3 Anterior segment of right lung [S3, superior lobe]
4 Anterior segment of left lung [S3, superior lobe]
5 Lateral basal segment of right lung [S9, inferior lobe]
6 Superior lingular segment of left lung [S4, superior lobe]
7 Medial basal segment of right lung [S7, inferior lobe]
8 Anterior basal segment of left lung [S8, inferior lobe]

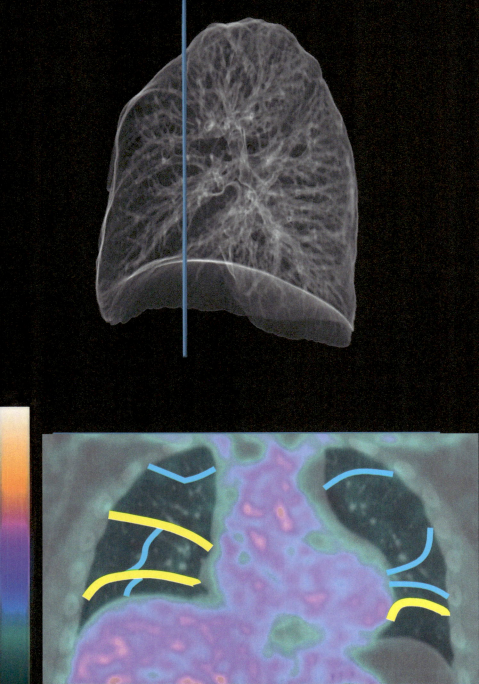

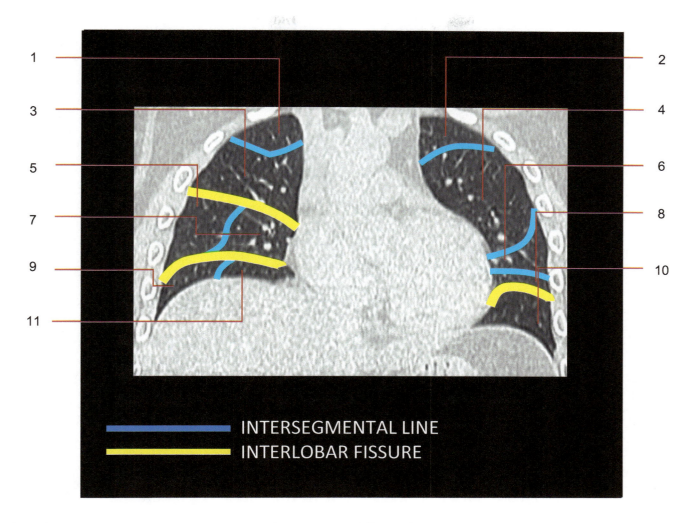

1 Apical segment of right lung [S1, superior lobe]
2 Apicoposterior segment of left lung [S1+2, superior lobe]
3 Anterior segment of right lung [S3, superior lobe]
4 Anterior segment of left lung [S3, superior lobe]
5 Lateral segment of right lung [S4, middle lobe]
6 Superior lingular segment of left lung [S4, superior lobe]
7 Medial segment of right lung [S5, middle lobe]
8 Inferior lingular segment of left lung [S5, superior lobe]
9 Anterior basal segment of right lung [S8, inferior lobe]
10 Anterior basal segment of left lung [S8, inferior lobe]
11 Medial basal segment of right lung [S7, inferior lobe]

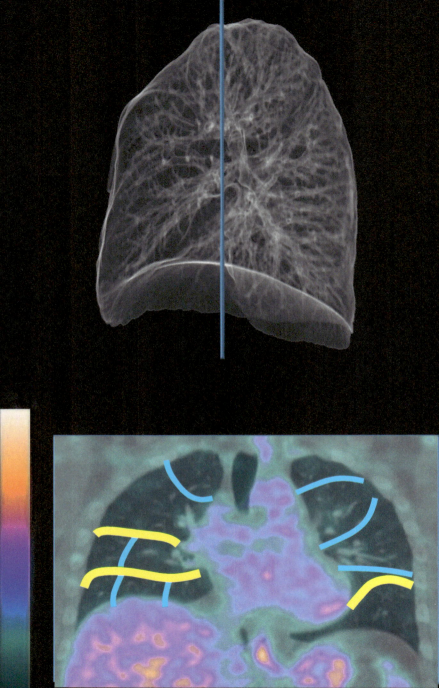

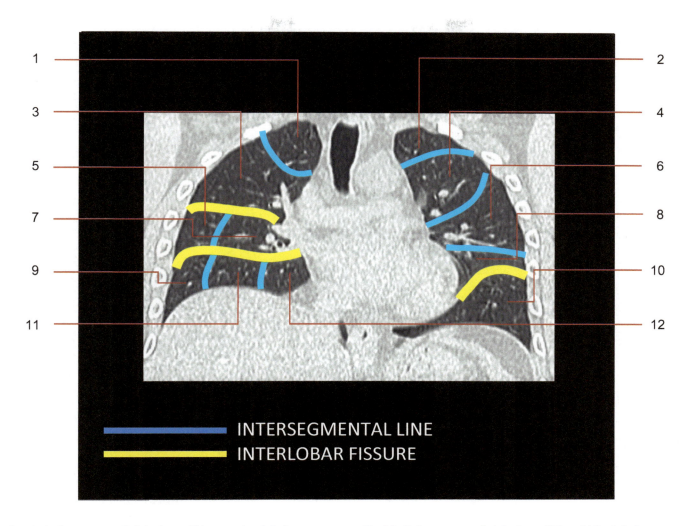

1 Apical segment of right lung [S1, superior lobe]
2 Apicoposterior segment of left lung [S1+2, superior lobe]
3 Anterior segment of right lung [S3, superior lobe]
4 Anterior segment of left lung [S3, superior lobe]
5 Lateral segment of right lung [S4, middle lobe]
6 Superior lingular segment of left lung [S4, superior lobe]
7 Medial segment of right lung [S5, middle lobe]
8 Inferior lingular segment of left lung [S5, superior lobe]
9 Anterior basal segment of right lung [S8, inferior lobe]
10 Anterior basal segment of left lung [S8, inferior lobe]
11 Lateral basal segment of right lung [S9, inferior lobe]
12 Medial basal segment of right lung [S7, inferior lobe]

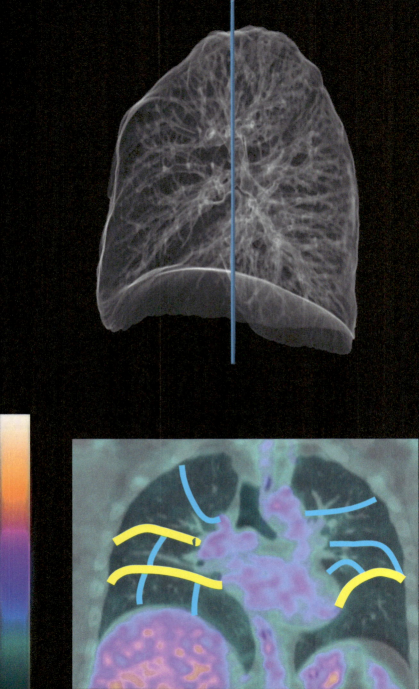

Thorax **Chapter** | 1   81

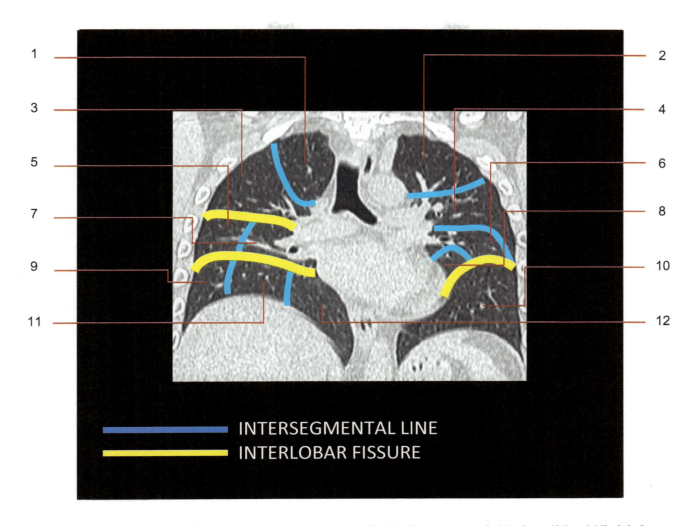

INTERSEGMENTAL LINE
INTERLOBAR FISSURE

1 Apical segment of right lung [S1, superior lobe]
2 Apicoposterior segment of left lung [S1+2, superior lobe]
3 Anterior segment of right lung [S3, superior lobe]
4 Anterior segment of left lung [S3, superior lobe]
5 Lateral segment of right lung [S4, middle lobe]
6 Superior lingular segment of left lung [S4, superior lobe]
7 Medial segment of right lung [S5, middle lobe]
8 Inferior lingular segment of left lung [S5, superior lobe]
9 Anterior basal segment of right lung [S8, inferior lobe]
10 Anterior basal segment of left lung [S8, inferior lobe]
11 Lateral basal segment of right lung [S9, inferior lobe]
12 Medial basal segment of right lung [S7, inferior lobe]

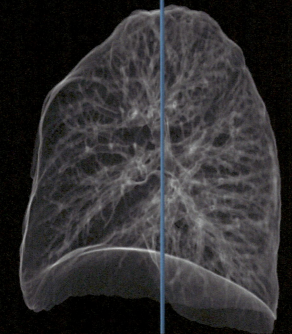

In coronal CT and PET/CT views, the identification of tracheal bifurcation and main bronchi can help in adequate assessment of pulmonary lobes and mediastinal lymph nodes [8].

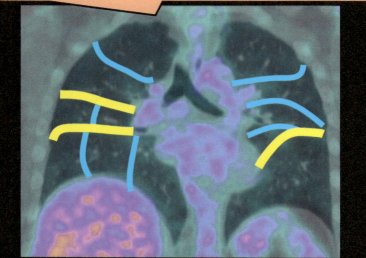

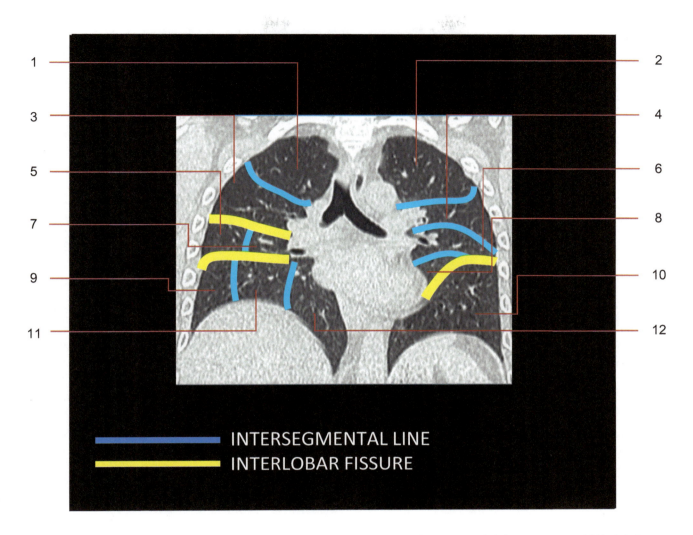

1 Apical segment of right lung [S1, superior lobe]
2 Apicoposterior segment of left lung [S1+2, superior lobe]
3 Anterior segment of right lung [S3, superior lobe]
4 Anterior segment of left lung [S3, superior lobe]
5 Lateral segment of right lung [S4, middle lobe]
6 Superior lingular segment of left lung [S4, superior lobe]
7 Medial segment of right lung [S5, middle lobe]
8 Inferior lingular segment of left lung [S5, superior lobe]
9 Anterior basal segment of right lung [S8, inferior lobe]
10 Anterior basal segment of left lung [S8, inferior lobe]
11 Lateral basal segment of right lung [S9, inferior lobe]
12 Medial basal segment of right lung [S7, inferior lobe]

See Ref [8]

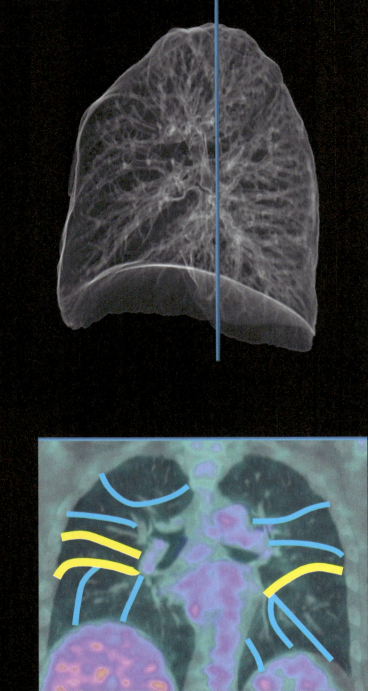

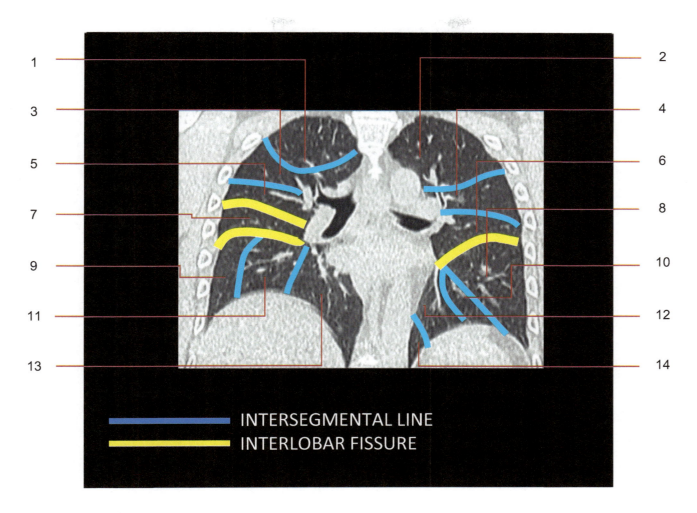

1 Apical segment of right lung [S1, superior lobe]
2 Apicoposterior segment of left lung [S1+2, superior lobe]
3 Posterior segment of right lung [S2, superior lobe]
4 Anterior segment of left lung [S3, superior lobe]
5 Anterior segment of right lung [S3, superior lobe]
6 Superior lingular segment of left lung [S4, superior lobe]
7 Lateral segment of right lung [S4, middle lobe]
8 Anterior basal segment of left lung [S8, inferior lobe]
9 Anterior basal segment of right lung [S8, inferior lobe]
10 Lateral basal segment of left lung [S9, inferior lobe]
11 Lateral basal segment of right lung [S9, inferior lobe]
12 Posterior basal segment of left lung [S10, inferior lobe]
13 Posterior basal segment of right lung [S10, inferior lobe]
14 Medial basal segment of left lung [S7, inferior lobe]

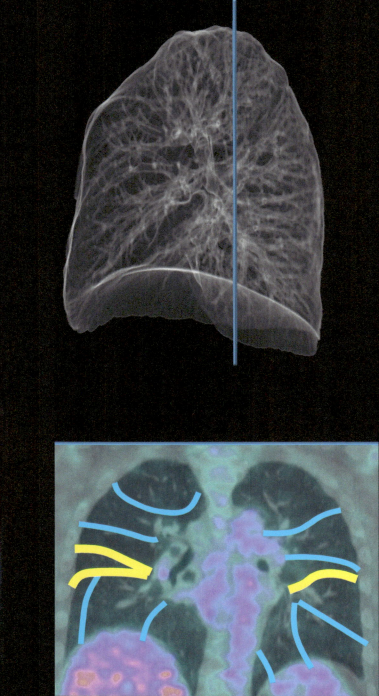

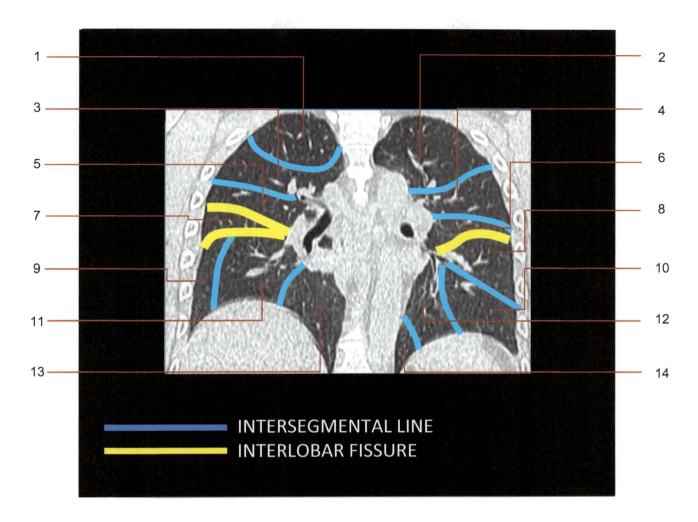

1 Apical segment of right lung [S1, superior lobe]
2 Apicoposterior segment of left lung [S1+2, superior lobe]
3 Posterior segment of right lung [S2, superior lobe]
4 Anterior segment of left lung [S3, superior lobe]
5 Anterior segment of right lung [S3, superior lobe]
6 Superior lingular segment of left lung [S4, superior lobe]
7 Lateral segment of right lung [S4, middle lobe]
8 Anterior basal segment of left lung [S8, inferior lobe]
9 Anterior basal segment of right lung [S8, inferior lobe]
10 Lateral basal segment of left lung [S9, inferior lobe]
11 Lateral basal segment of right lung [S9, inferior lobe]
12 Posterior basal segment of left lung [S10, inferior lobe]
13 Posterior basal segment of right lung [S10, inferior lobe]
14 Medial basal segment of left lung [S7, inferior lobe]

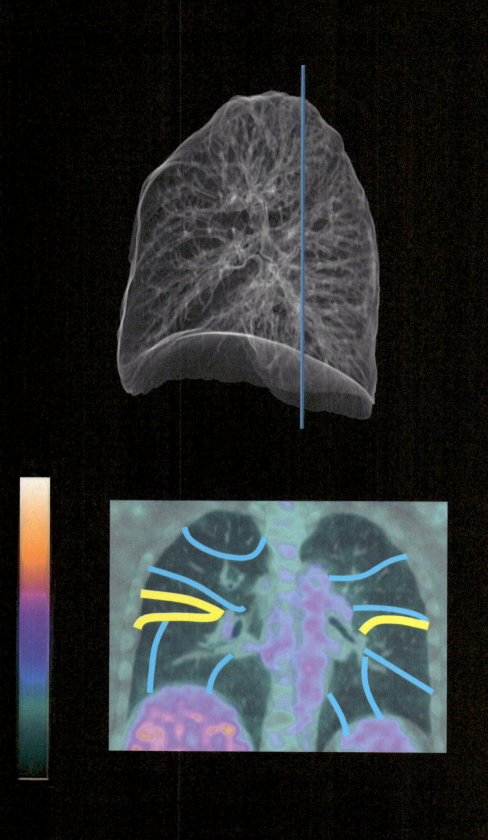

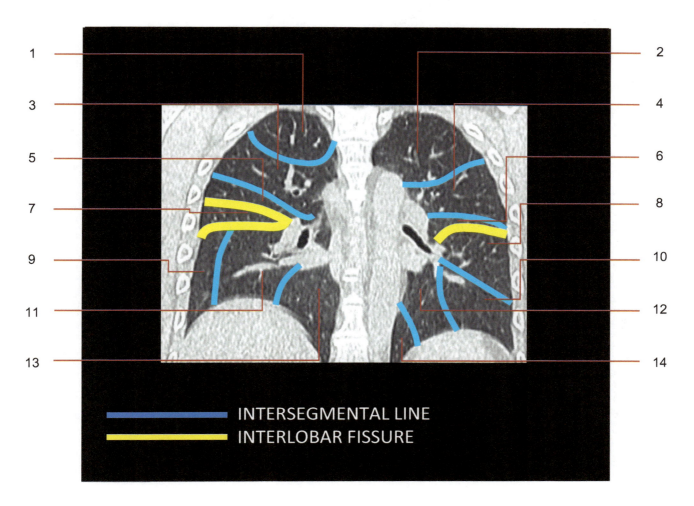

1 Apical segment of right lung [S1, superior lobe]
2 Apicoposterior segment of left lung [S1+2, superior lobe]
3 Posterior segment of right lung [S2, superior lobe]
4 Anterior segment of left lung [S3, superior lobe]
5 Anterior segment of right lung [S3, superior lobe]
6 Superior lingular segment of left lung [S4, superior lobe]
7 Lateral segment of right lung [S4, middle lobe]
8 Anterior basal segment of left lung [S8, inferior lobe]
9 Anterior basal segment of right lung [S8, inferior lobe]
10 Lateral basal segment of left lung [S9, inferior lobe]
11 Lateral basal segment of right lung [S9, inferior lobe]
12 Posterior basal segment of left lung [S10, inferior lobe]
13 Posterior basal segment of right lung [S10, inferior lobe]
14 Medial basal segment of left lung [S7, inferior lobe]

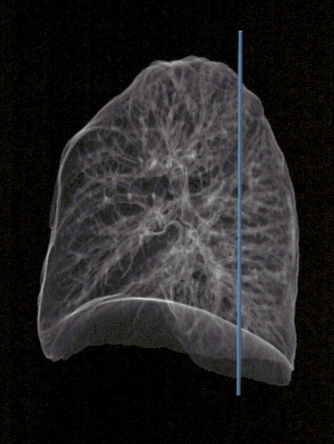
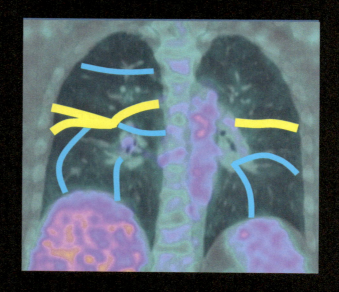

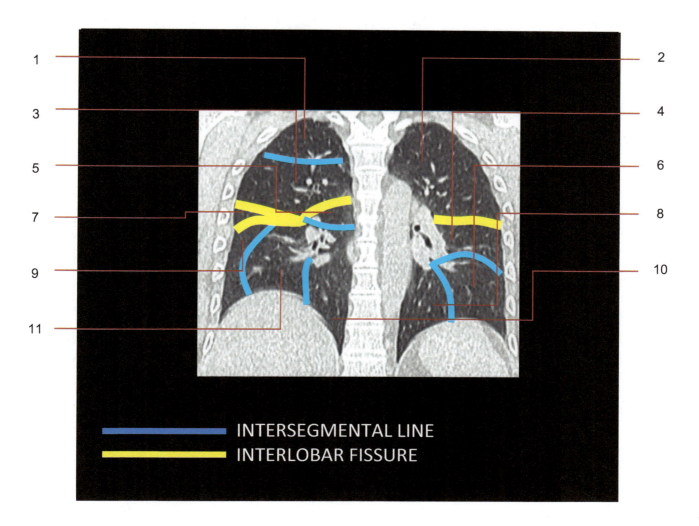

1 Apical segment of right lung [S1, superior lobe]
2 Apicoposterior segment of left lung [S1+2, superior lobe]
3 Posterior segment of right lung [S2, superior lobe]
4 Superior segment of left lung [S6, inferior lobe]
5 Superior segment of right lung [S6, inferior lobe]
6 Lateral basal segment of left lung [S9, inferior lobe]
7 Lateral segment of right lung [S4, middle lobe]
8 Posterior basal segment of left lung [S10, inferior lobe]
9 Anterior basal segment of right lung [S8, inferior lobe]
10 Posterior basal segment of right lung [S10, inferior lobe]
11 Lateral basal segment of right lung [S9, inferior lobe]

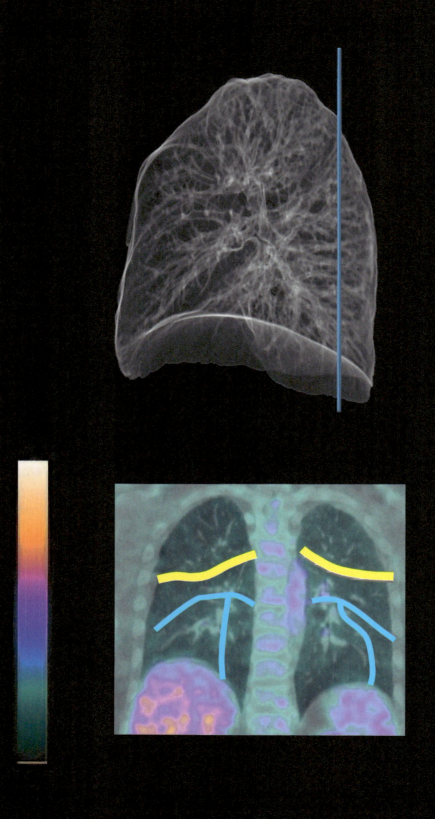

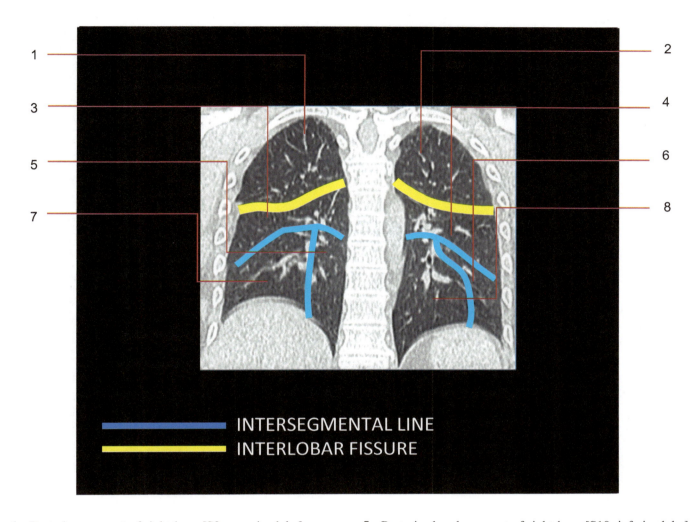

1 Posterior segment of right lung [S2, superior lobe]
2 Apicoposterior segment of left lung [S1+2, superior lobe]
3 Superior segment of right lung [S6, inferior lobe]
4 Superior segment of left lung [S6, inferior lobe]
5 Posterior basal segment of right lung [S10, inferior lobe]
6 Lateral basal segment of left lung [S9, inferior lobe]
7 Lateral basal segment of right lung [S9, inferior lobe]
8 Posterior basal segment of left lung [S10, inferior lobe]

1.2 Mediastinum PET/CT
1.2.1 Anatomy

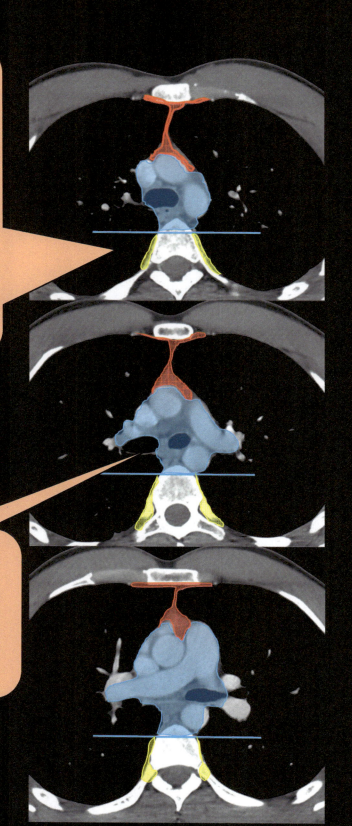

Several classification schemes for the mediastinum have been created and used to varying degrees in clinical practice. Anatomists divide the mediastinum into four parts: superior, anterior inferior, middle inferior, posterior inferior mediastinum [9].

The International Thymic Malignancy Interest Group (ITMIG) has introduced a new definition of mediastinal compartments to be used with cross-sectional imaging and adopted as a new standard [10].

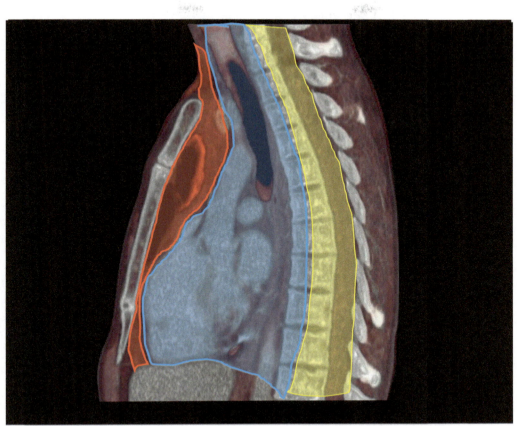

International Thymic Malignancy Interest Group (ITMIG) classification system for mediastinal compartments

PREVASCULAR	Superior: thoracic inlet Inferior: diaphragm Anterior: sternum Lateral: parietal mediastinal pleura Posterior: anterior aspect of the pericardium as it wraps around the heart in a curvilinear fashion
VISCERAL	Superior: thoracic inlet Inferior: diaphragm Anterior: posterior boundaries of the prevascular compartment Posterior: vertical line connecting a point on each thoracic vertebral body 1 cm posterior to its anterior margin
PARAVERTEBRAL	Superior: thoracic inlet Inferior: diaphragm Anterior: posterior boundaries of the visceral compartment Posterolateral: vertical line against the posterior margin of the chest wall at the lateral margin of the transverse process of the thoracic spine

See Refs [9,10]

1.2.2 Axial

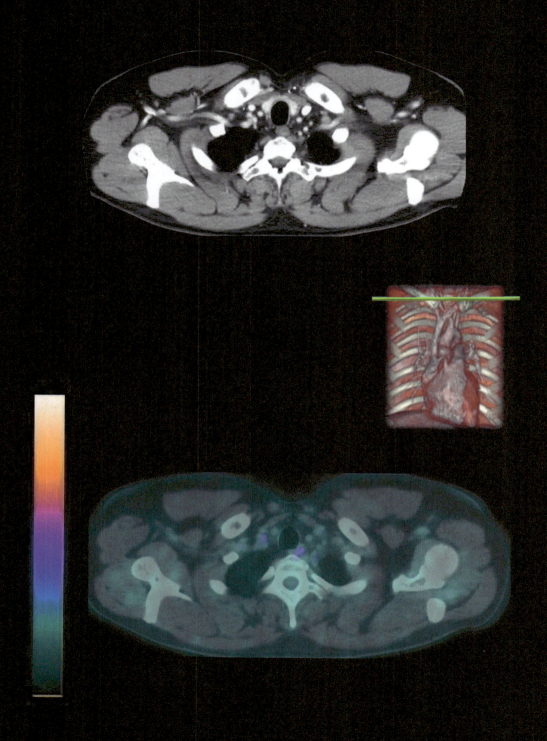

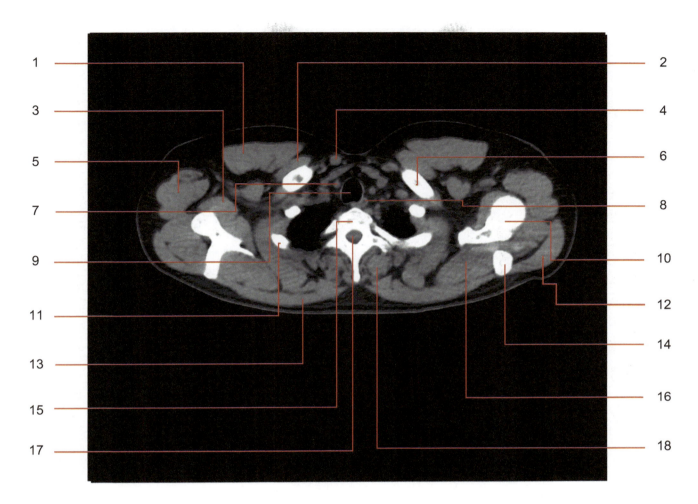

1	Pectoralis major	10	Scapula
2	Subclavius muscle	11	Rib
3	Subscapularis muscle	12	Deltoid (spinal)
4	Sternocleidomastoid	13	Trapezius muscle
5	Deltoid (clavicular)	14	Scapula (spine)
6	Clavicle	15	Dorsal vertebra
7	Right carotid artery	16	Supraspinatus muscle
8	Esophagus	17	Spinal cord
9	Trachea	18	Erector spinae muscle

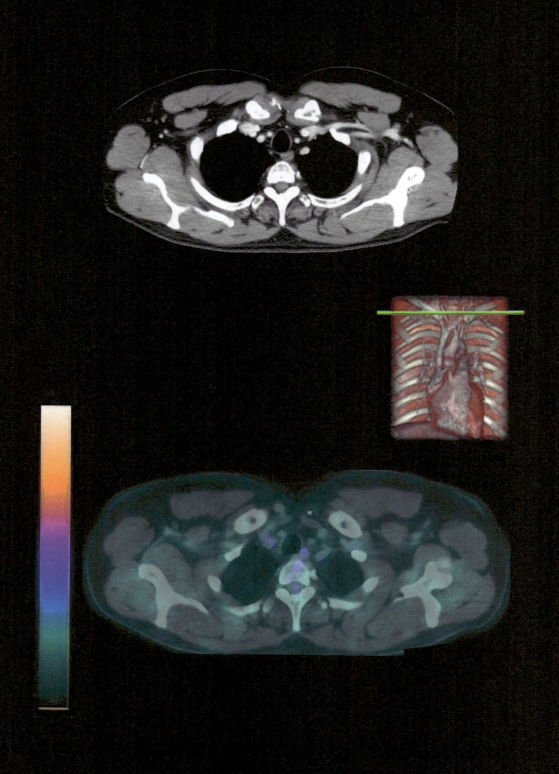

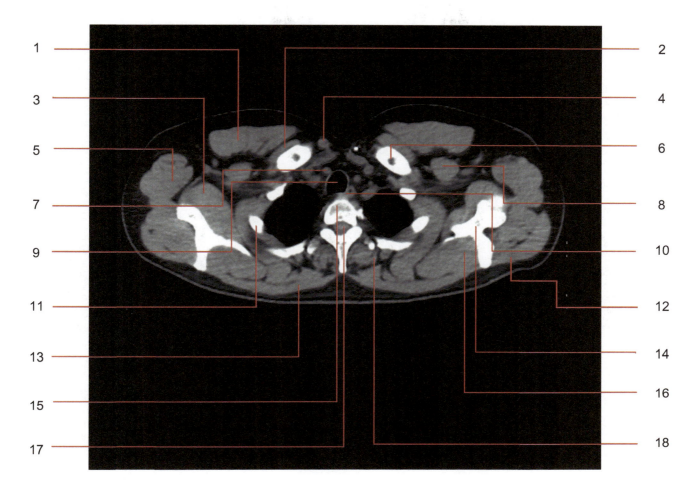

1 Pectoralis major
2 Subclavius muscle
3 Subscapularis muscle
4 Sternocleidomastoid
5 Deltoid (clavicular)
6 Clavicle
7 Right carotid artery
8 Pectoralis minor
9 Trachea
10 Esophagus
11 Rib
12 Deltoid (spinal)
13 Trapezius muscle
14 Scapula
15 Dorsal vertebra
16 Supraspinatus muscle
17 Spinal cord
18 Erector spinae muscle

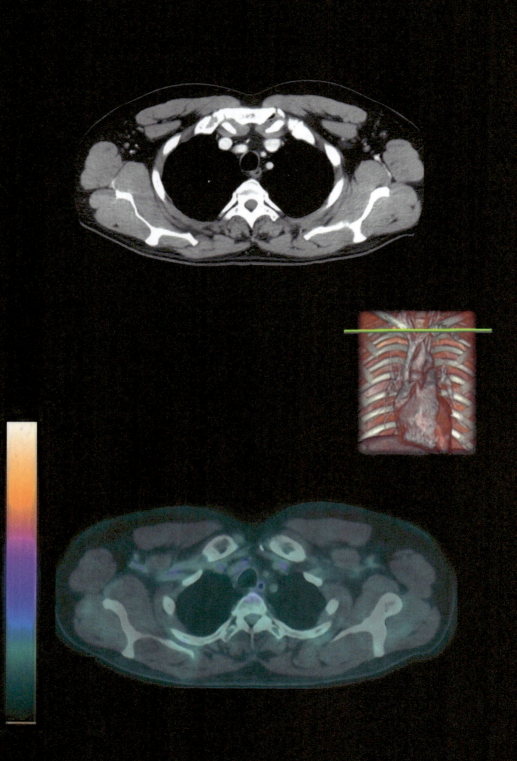

Thorax Chapter | 1   101

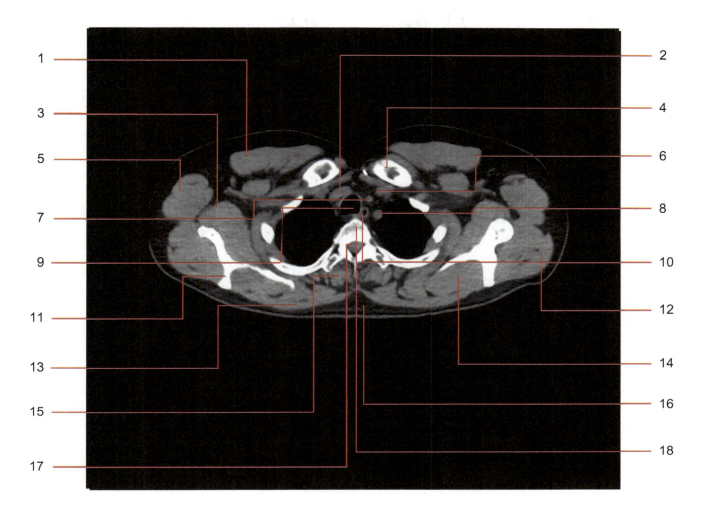

1	Pectoralis major	10	Esophagus
2	Brachiocephalic vein	11	Scapula
3	Subscapularis muscle	12	Deltoid (spinal)
4	Clavicle	13	Trapezius muscle
5	Deltoid (clavicular)	14	Supraspinatus muscle
6	Left subclavian vein	15	Erector spinae muscle
7	Left carotid artery	16	Dorsal subcutaneous adipose tissue
8	Left subclavian artery	17	Spinal cord
9	Trachea	18	Dorsal vertebra

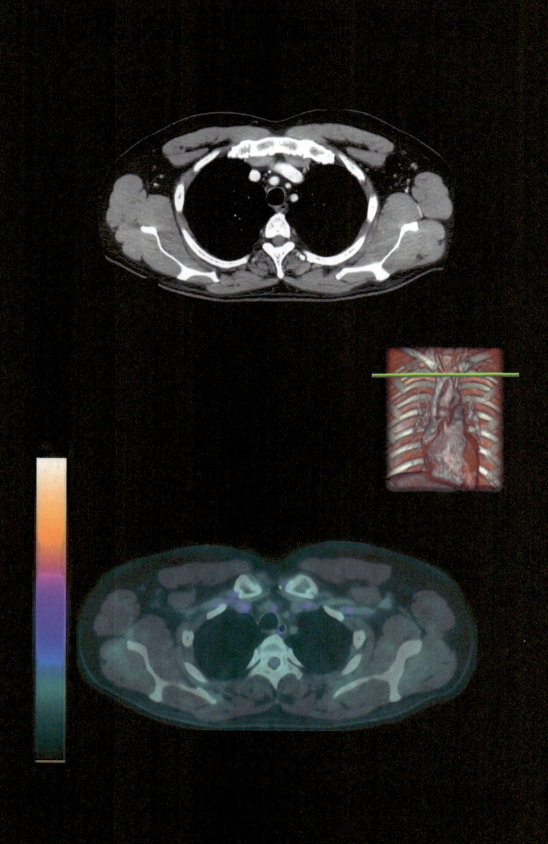

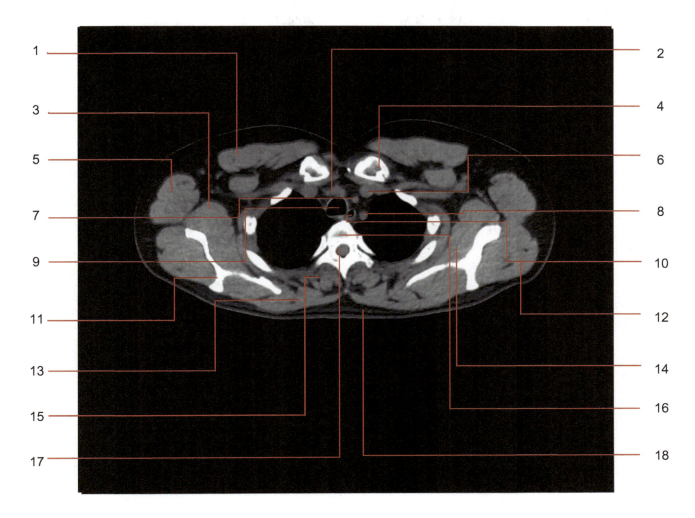

1 Pectoralis major
2 Right brachiocephalic vein
3 Subscapularis muscle
4 Clavicle
5 Deltoid (clavicular)
6 Left subclavian vein
7 Left carotid artery
8 Left subclavian artery
9 Trachea
10 Esophagus
11 Scapula
12 Deltoid (Spinal)
13 Trapezius muscle
14 Supscapularis muscle
15 Erector spinae muscle
16 Dorsal vertebra
17 Spinal cord
18 Dorsal subcutaneous adipose tissue

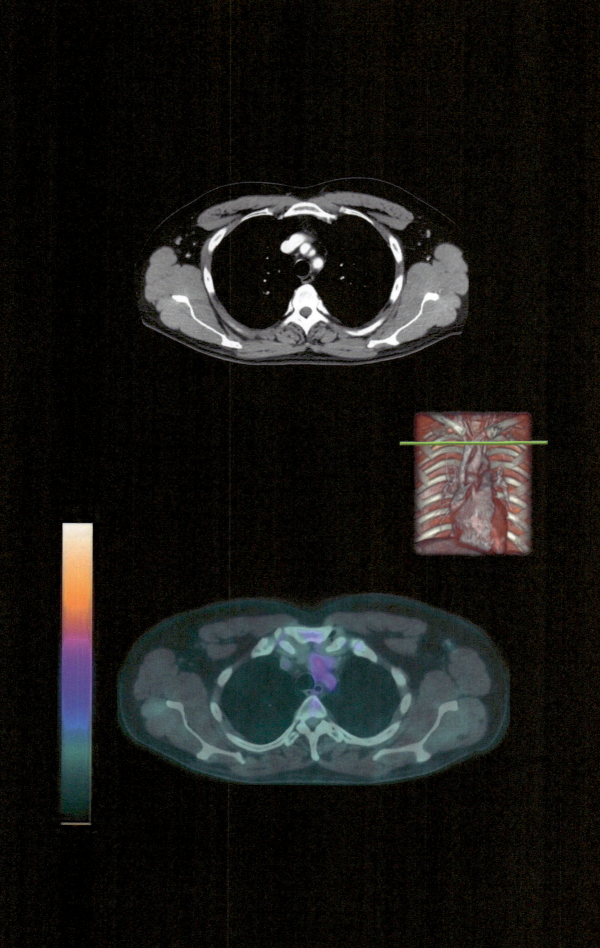

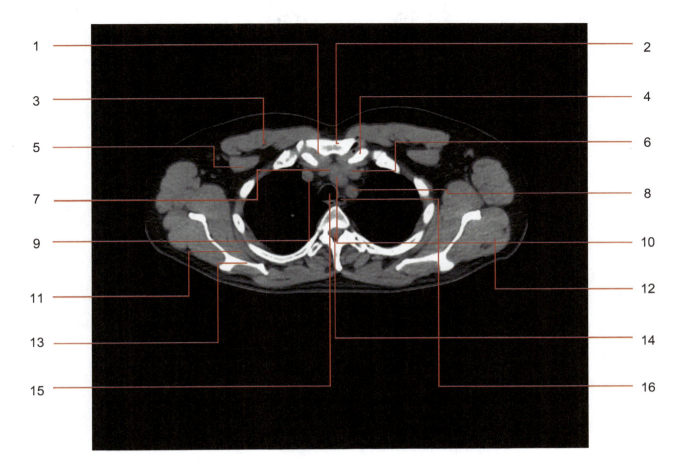

1	Sternoclavicular joint	9	Right brachiocephalic vein
2	Sternum (manubrium)	10	Brachiocephalic artery
3	Pectoralis major	11	Subscapularis muscle
4	Clavicle	12	Deltoid (spinal)
5	Pectoralis minor	13	Scapula
6	Left subclavian vein	14	Spinal cord
7	Left brachiocephalic vein	15	Trachea
8	Left subclavian artery	16	Esophagus

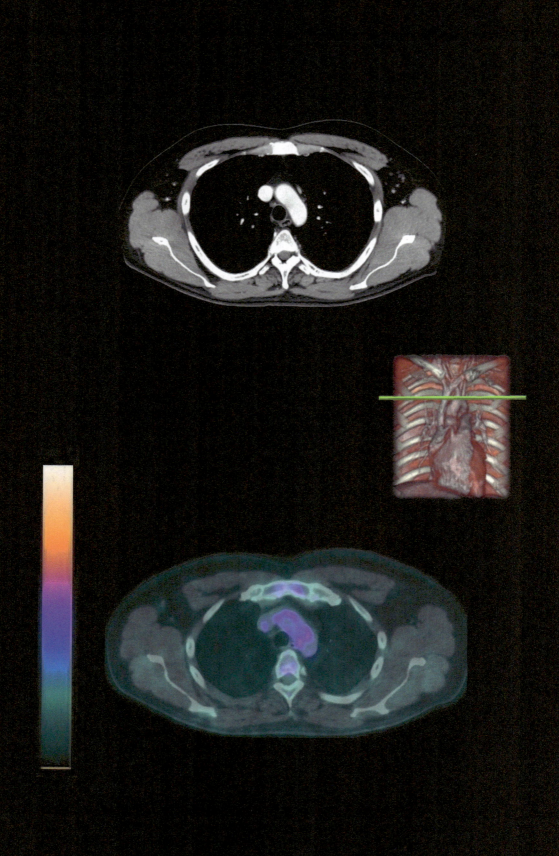

Thorax Chapter | 1   107

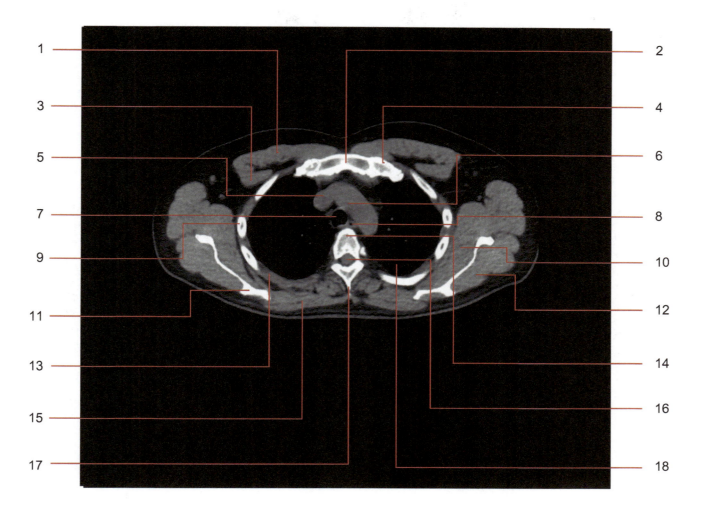

1	Pectoralis major	10	Subscapularis muscle
2	Sternum (manubrium)	11	Scapula
3	Pectoralis minor	12	Infraspinous muscle
4	Clavicle	13	Intercostal muscle
5	Superior vena cava	14	Dorsal vertebra (body)
6	Aortic arch	15	Trapezius muscle
7	Trachea	16	Spinal cord
8	Esophagus	17	Dorsal vertebra (spinous process)
9	Rib	18	Lung

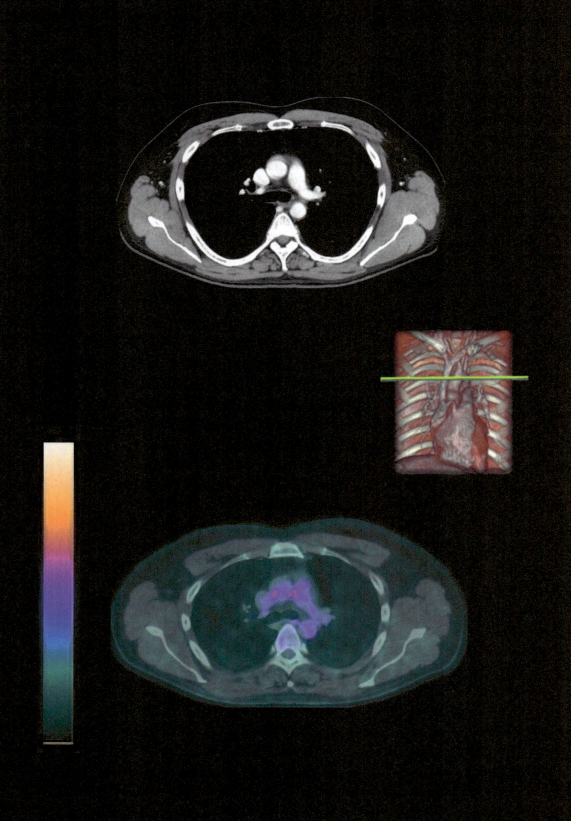

Thorax Chapter | 1   109

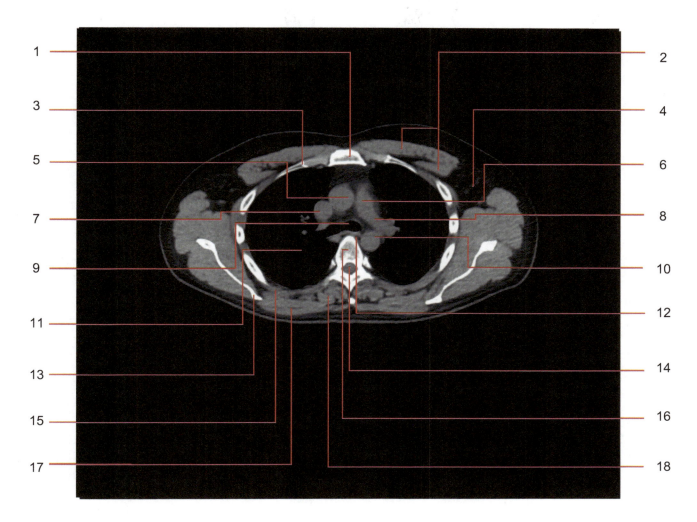

1 Sternum (manubrium)
2 Pectoralis major and pectoralis minor
3 Costochondral joint
4 Axilla
5 Ascending aorta
6 Pulmonary artery
7 Superior vena cava
8 Left pulmonary artery
9 Trachea (carina)
10 Descending aorta
11 Lung
12 Esophagus
13 Scapula
14 Spinal cord
15 Intercostal muscle
16 Dorsal vertebra (body)
17 Trapezius muscle
18 Erector spinae muscle

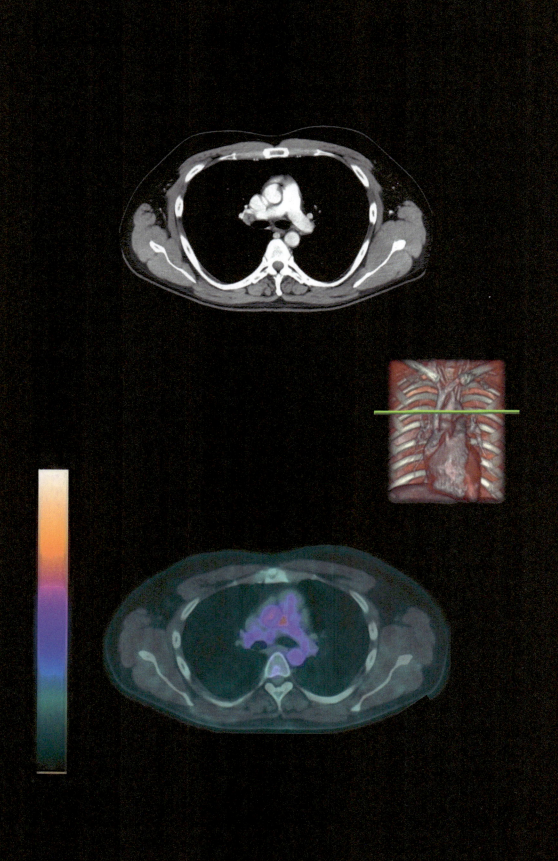

Thorax Chapter | 1 111

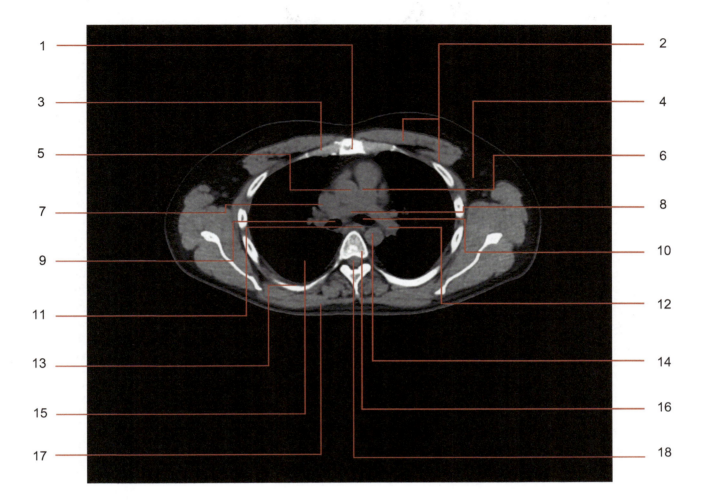

1 Manubriosternal joint
2 Pectoralis major and pectoralis minor
3 Costochondral joint
4 Axilla
5 Ascending aorta
6 Pulmonary artery
7 Superior vena cava
8 Right pulmonary artery
9 Right main bronchus
10 Left main bronchus
11 Esophagus
12 Left pulmonary artery
13 Rib
14 Descending aorta
15 Lung
16 Dorsal vertebra (body)
17 Trapezius muscle
18 Spinal cord

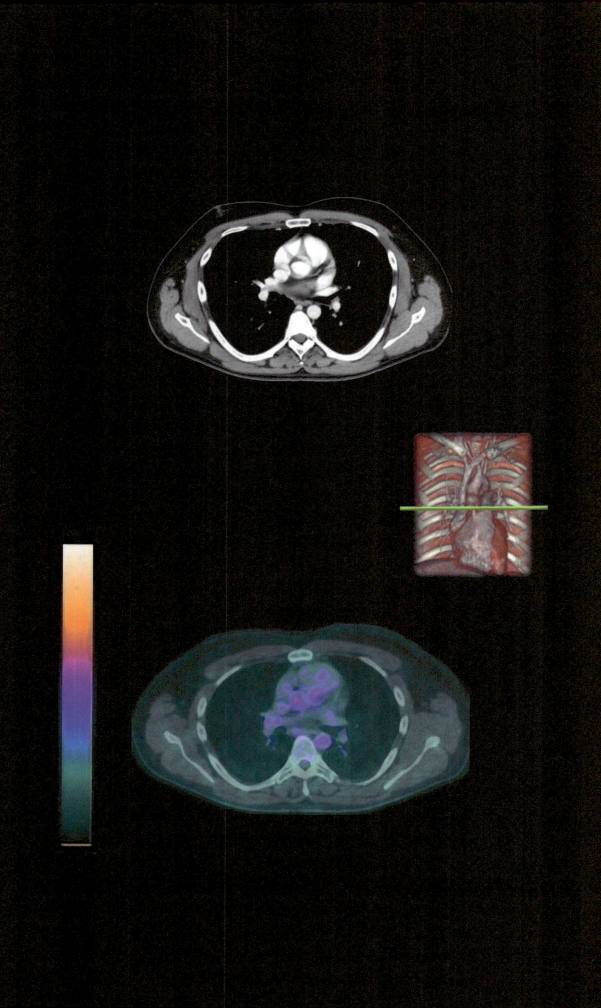

Thorax Chapter | 1   113

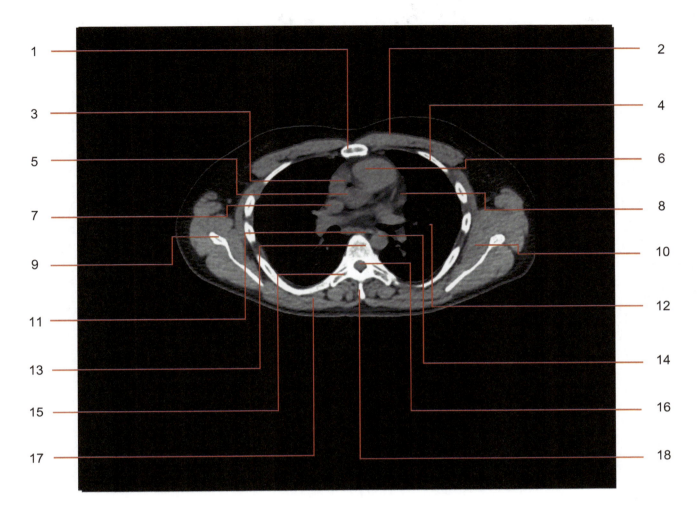

1	Sternum (body)	10	Subscapularis muscle
2	Pectoralis major	11	Esophagus
3	Right auricle	12	Lung
4	Rib	13	Dorsal vertebra (body)
5	Ascending aorta	14	Descending aorta
6	Right ventricle	15	Vertebra (transverse process)
7	Superior vena cava	16	Spine
8	Left auricle	17	Trapezius muscle
9	Scapula	18	Dorsal vertebra (spinous process)

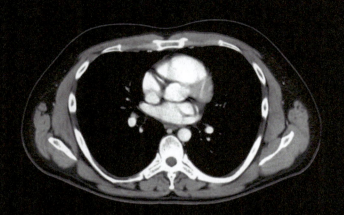

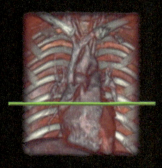

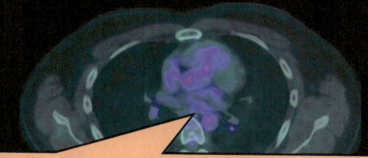

Visual analysis of mediastinal *blood pool* activity in PET/CT fusion images is of the utmost importance in the assessment of malignant lesions [11] or evaluation of response to therapy in lymphomas as referred to *interim PET* in Hodgkin's lymphoma. [12].

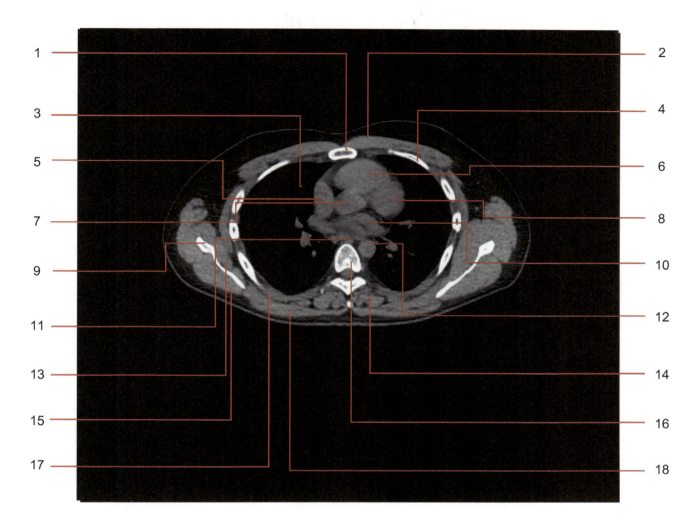

1	Sternum (body)	10	Left atrium
2	Pectoralis major	11	Esophagus
3	Lung	12	Descending aorta
4	Rib	13	Subscapularis muscle
5	Right atrium	14	Erector spinae muscle
6	Right ventricle	15	Infraspinous muscle
7	Aortic bulb	16	Dorsal vertebra
8	Left ventricle	17	Intercostal muscle
9	Scapula	18	Trapezius muscle

See Refs [11,12]

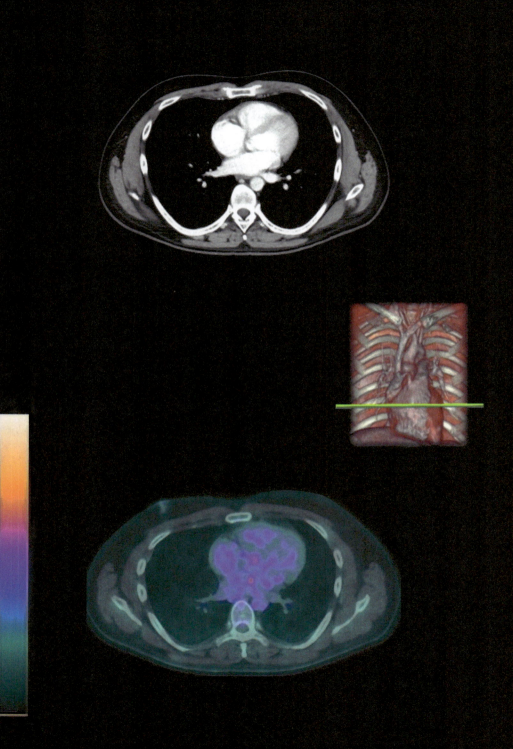

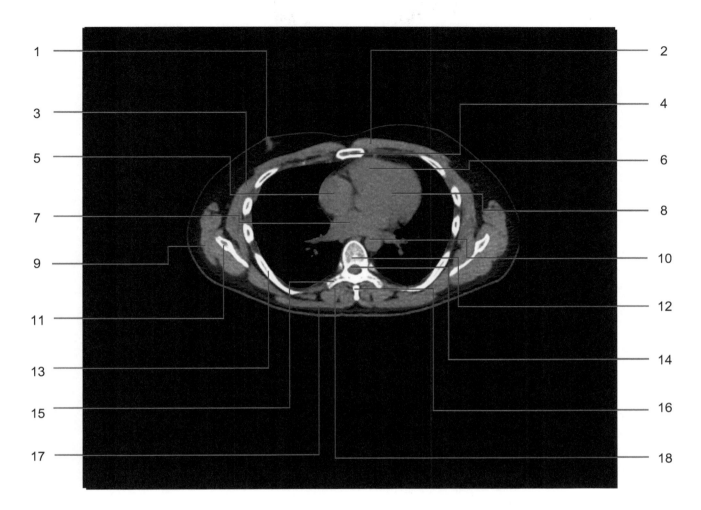

1	Right nipple	10	Descending aorta
2	Pectoralis major	11	Scapula
3	Serratus anterior muscle	12	Dorsal vertebra (body)
4	Sternum (body)	13	Rib
5	Right atrium	14	Spinal cord
6	Right ventricle	15	Vertebra (transverse process)
7	Left atrium	16	Vertebra (spinous process)
8	Left ventricle	17	Trapezius muscle
9	Latissimus dorsi	18	Erector spinae muscle

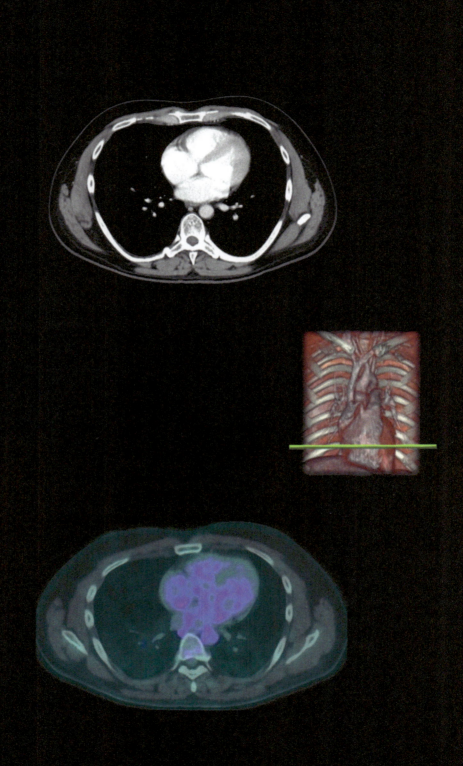

Thorax Chapter | 1   119

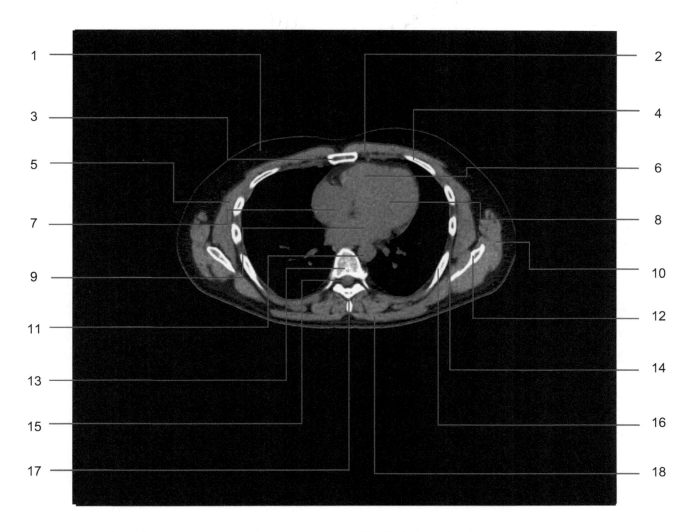

1   Subcutaneous adipose tissue
2   Pectoralis major
3   Sternum (body)
4   Rib
5   Serratus anterior muscle
6   Right ventricle
7   Right atrium
8   Left ventricle
9   Left atrium
10 Latissimus dorsi
11 Descending aorta
12 Scapula
13 Dorsal vertebra (body)
14 Intercostal muscle
15 Spinal cord
16 Rib
17 Vertebra (spinous process)
18 Trapezius

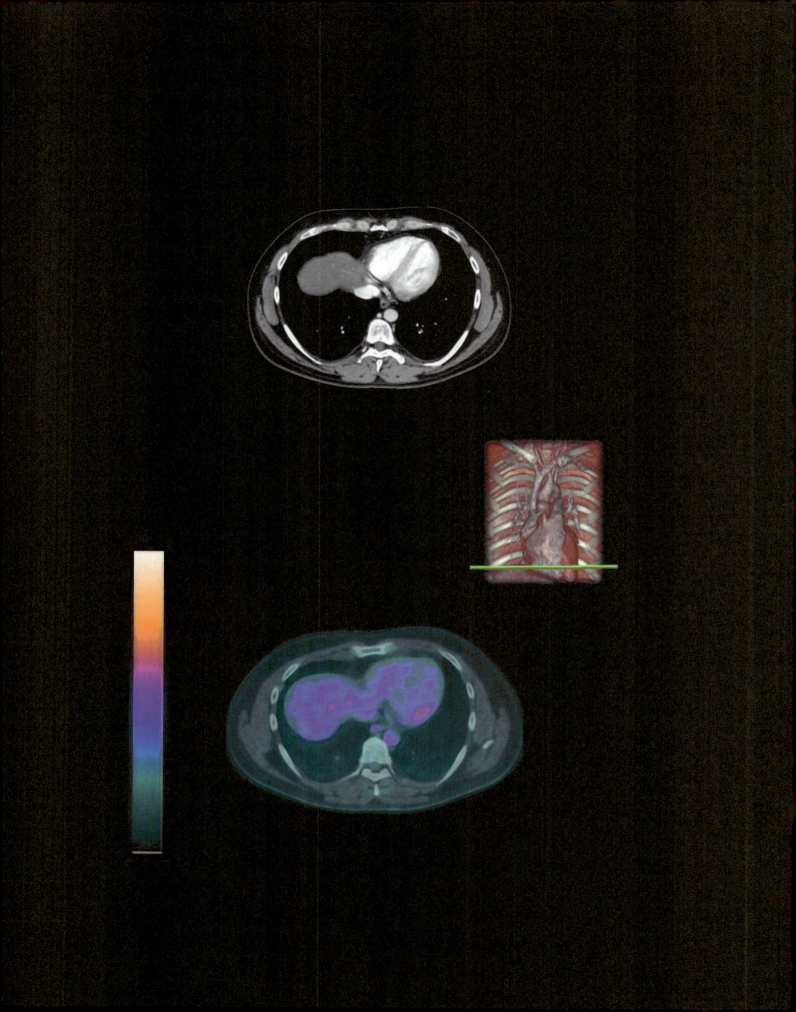

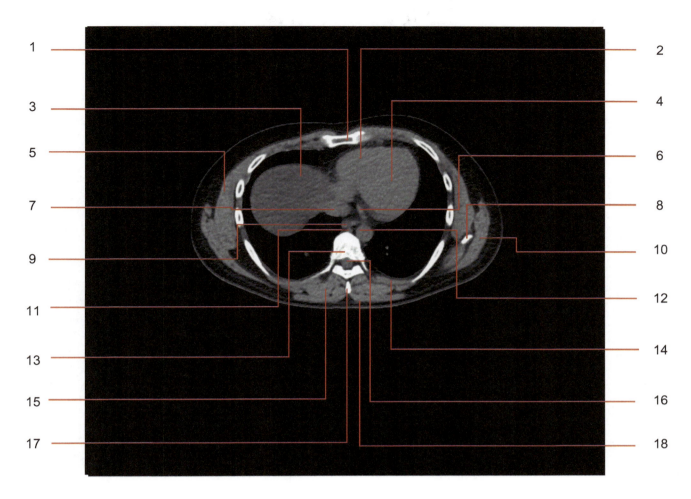

1 Sternum (body)
2 Right ventricle
3 Liver
4 Left ventricle
5 Serratus anterior muscle
6 Coronary sinus
7 Inferior vena cava
8 Scapula (inferior angle)
9 Esophagus
10 Latissimus dorsi
11 Azygos vein
12 Descending aorta
13 Dorsal vertebra (body)
14 Intercostal muscle
15 Erector spinae muscle
16 Spinal cord
17 Vertebra (spinous process)
18 Subcutaneous adipose tissue

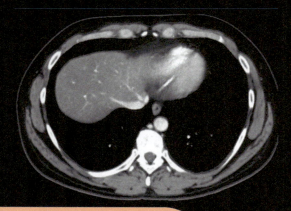

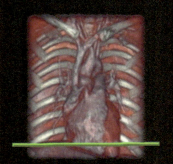

Liver ¹⁸F-FDG uptake is relatively stable. Its visual assessment in PET/CT fusion images is important as visual cut-off in evaluation of malignant lesions [13].
Tracer injection during non fasting state determines higher ¹⁸F-FDG uptake in liver and muscles, potentially enabling evaluation of PET data [14].

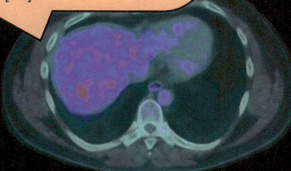

Thorax **Chapter** | 1 123

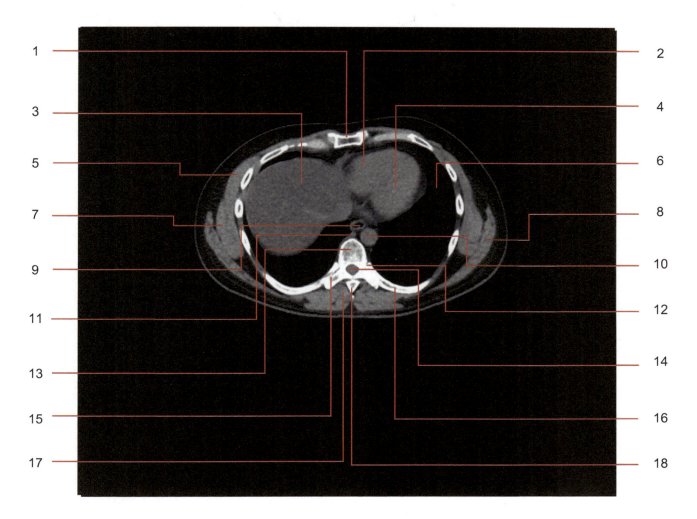

1 Sternum (body)
2 Right ventricle
3 Liver
4 Left ventricle
5 External oblique muscle
6 Lung
7 Serratus anterior muscle
8 Latissimus dorsi
9 Esophagus
10 Descending aorta
11 Azygos vein
12 Costovertebral joint
13 Dorsal vertebra (body)
14 Spinal cord
15 Vertebra (transverse process)
16 Rib
17 Erector spinae muscle
18 Vertebra (spinous process)

See Refs [13,14]

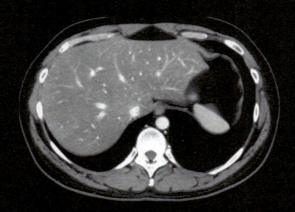
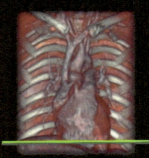

In some conditions as cachexia or severe dismetabolic sindromes the ^{18}F-FDG uptake in liver can considerably decrease. Cancer patients with extremely decreased liver FDG uptake are likely to have lower overall survival [15].

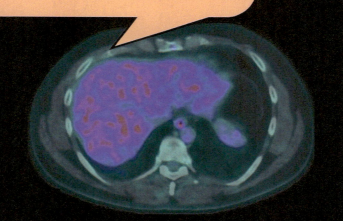

Thorax Chapter | 1   125

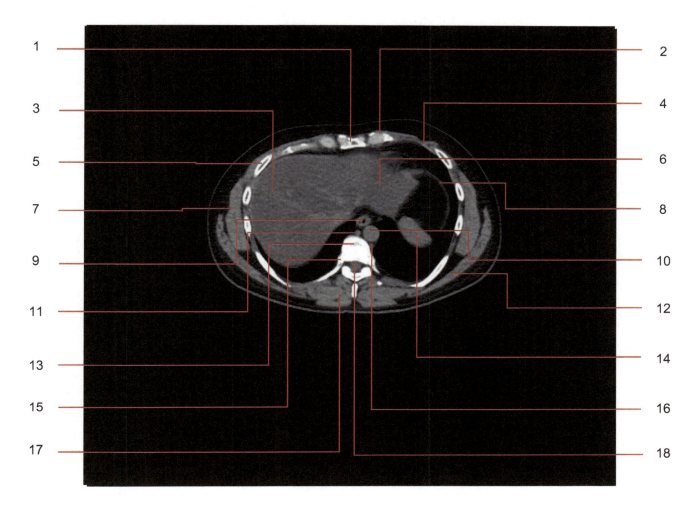

1	Sternum (body)	10	Descending aorta
2	Sternocostal joint	11	Azygos vein
3	Right hepatic lobe	12	Latissimus dorsi
4	Intercostal muscle	13	Dorsal vertebra (body)
5	Rib	14	Spleen (upper pole)
6	Left hepatic lobe	15	Costovertebral joint
7	External oblique muscle	16	Hemiazygos vein
8	Diaphragm	17	Erector spinae muscle
9	Esophagus	18	Spinal cord

See Ref [15]

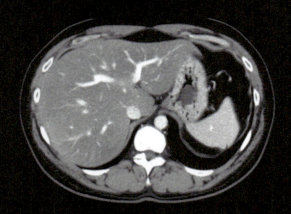

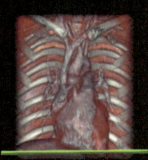

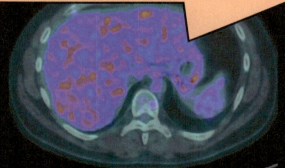

¹⁸F-FDG uptake in the stomach can be linked to normal activity of gastric mucosa [16].

Thorax Chapter | 1   127

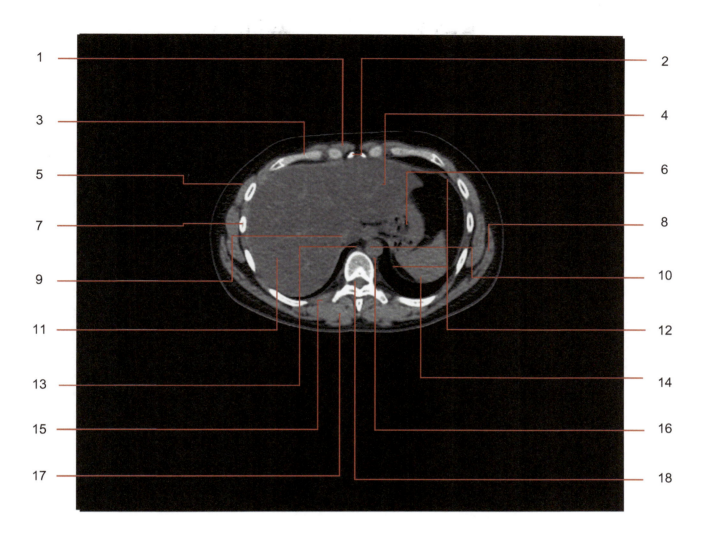

1 Rectus abdominis muscle
2 Xiphoid process
3 Rib (costal cartilage)
4 Left hepatic lobe
5 External oblique muscle
6 Stomach (fundus)
7 Rib
8 Latissimus dorsi
9 Inferior vena cava (retrohepatic tract)
10 Descending aorta
11 Right hepatic lobe
12 Diaphragm
13 Azygos vein
14 Spleen
15 Intercostal muscle
16 Hemiazygos vein
17 Erector spinae muscle
18 Spinal cord

See Ref [16]

128 Atlas of Hybrid Imaging

1.3 Clinical cases, tricks, and pitfalls
1.3.1 ^{18}F-FDG

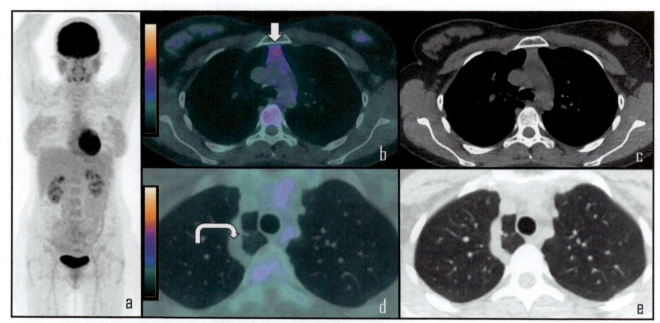

FIGURE 1.1 A 33-year-old woman was examined by means of ^{18}F-FDG PET/CT during follow-up of breast cancer, one month following chemotherapy. No foci of pathologic tracer uptake were detected in whole body PET *Maximum Intensity Projection* (a). Axial PET/CT view showed mild ^{18}F-FDG uptake in the upper mediastinum (b, *arrow*), corresponding to slightly hypodense tissue in correlative CT view (c). This finding was considered as residual thymic activity following chemotherapy [17]. In the same patient, an accessory azygos lobe was detected in the upper lobe of the right lung, due to anomalous lateral course of the azygos vein [18], as showed in axial PET/CT (d, *curved arrow*) and CT (e) views with lung window level.

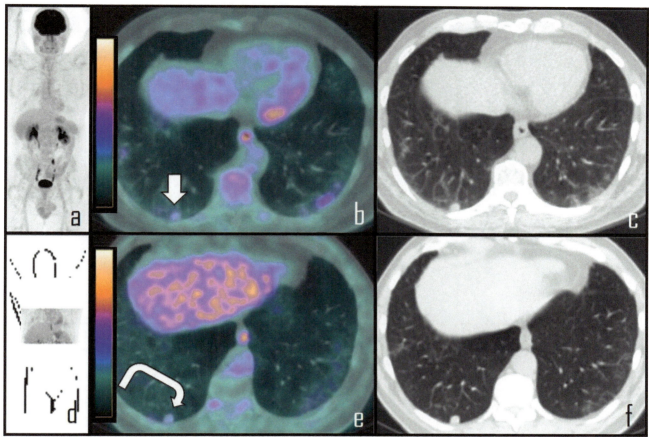

FIGURE 1.2 A 56-year-old male patient was examined for characterization of a solitary node in the inferior lobe of the right lung. ^{18}F-FDG PET *Maximum Intensity Projection* (a) was negative. Axial PET/CT (b, *arrow*) and CT (c) views confirmed a solitary lung node in the right inferior lobe, without significant metabolic activity. Respiratory 4D gated PET/CT of the inferior thorax, performed after whole body scan (d), better displays a 1 cm wide node without significant glucidic activity in PET/CT, higher signal to background quality fusion images (e, *curved arrow*), regular margins and better CT resolution (f).

130 Atlas of Hybrid Imaging

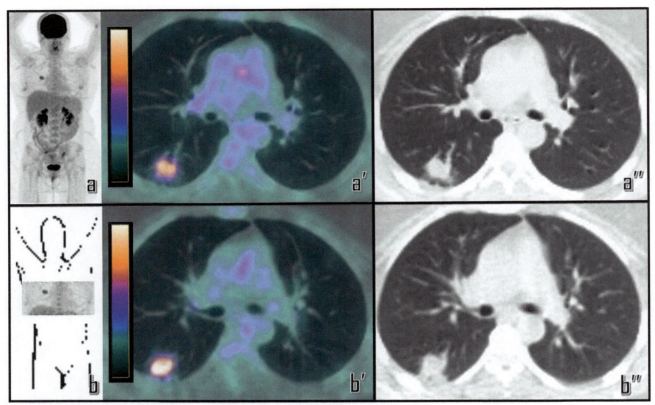

FIGURE 1.3 Metabolic characterization of a solitary lung node with *dual-phase* ^{18}F-FDG PET/CT. Whole body PET (a), 60 minutes following ^{18}F-FDG administration, shows intense tracer uptake in the right lung, with SUVmax 4.5, corresponding to a lung node with irregular margins of the apical segment of the right inferior lobe, as evident in correlative axial PET/CT (a') and CT (a'') views. Late PET/CT scan of the thorax (b), performed 120 minutes' after injection, displayed rise of the uptake in the lesion, with SUVmax 6.2, evident in correlative PET/CT (b') and CT (b'') views. Histological exam diagnosed squamous carcinoma. By means of *dual-phase* PET/CT, malignant lesions generally show rise of ^{18}F-FDG uptake during the time, due to upregulation of glucose transporters in cancer cells. This technique can lead a better confidence in diagnosing suspicious malignant pulmonary lesions [19].

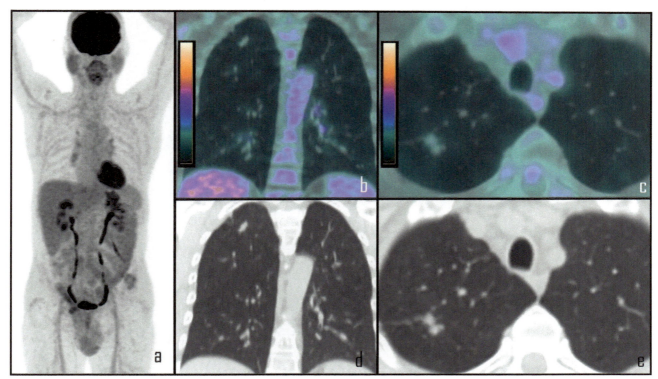

FIGURE 1.4 A 71-year-old patient was examined by means of ^{18}F-FDG PET/CT for characterization of a solitary lung node in the apex of the right lung. PET (a) was negative. Coronal (b) and axial (c) PET/CT did not show ^{18}F-FDG uptake in the node. Nevertheless, correlative coronal (d) and axial (e) CT views showed this finding as bilobed, with irregular margins, 2 cm wide. Due to morphological features, this node was submitted to biopsy. Histological exam diagnosed bronchioloalveolar carcinoma. Despite low glycolytic activity, some lung lesions can show malignant behavior. Knowledge of CT diagnostic criteria in hybrid imaging is of the utmost importance in identifying these false negative cases.

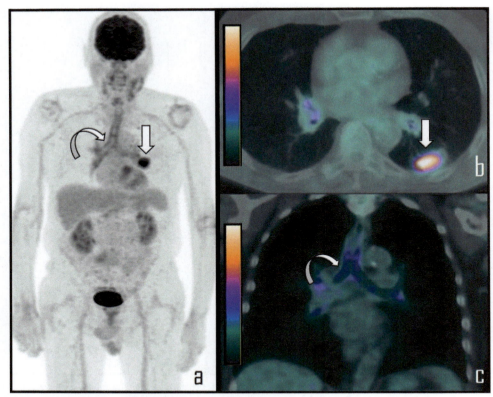

FIGURE 1.5 Whole body ^{18}F-FDG PET *Maximum Intensity Projection* (a) of a 63-year-old patient, in staging of lung cancer, showing pathologic tracer uptake in left lung (*arrow*) and diffuse, mild uptake in main bronchi and trachea (*curved arrow*). Axial PET/CT (b, *arrow*) confirms the a 3-cm-wide neoplastic pulmonary lesion in the superior segment of the left inferior lobe (SUVmax 8.7); coronal PET/CT view (c) also confirms diffuse, mild ^{18}F-FDG uptake in the tracheobronchial district (*curved arrow*), due to inflammation. Patient anamnesis was positive for cough in the last two weeks, with mucus production. Being a condition of inflammation, tracheobronchitis can present ^{18}F-FDG uptake, due to activation of white cells as monocytes and macrophages [20].

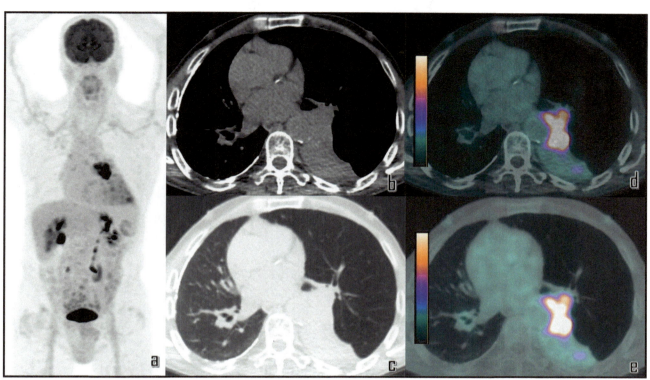

FIGURE 1.6 Whole body ^{18}F-FDG PET *Maximum Intensity Projection* (a) of a 72-year-old patient, during staging of lung cancer, showing pathologic tracer uptake in left lung. Axial CT with mediastinal (b) and lung parenchyma (c) window levels confirm extended area of consolidation in left lung, with nonhomogeneous density and irregular margins. Nevertheless, correlative axial PET/CT views (d, e) show the metabolically active component of the lesion, allowing to identify atelectasia downstream of the bronchial neoplastic stenosis. This feature in hybrid *PET/CT era* allows to adequately plan surgery and/or radiotherapy, aimed to preserve normal lung parenchyma [21].

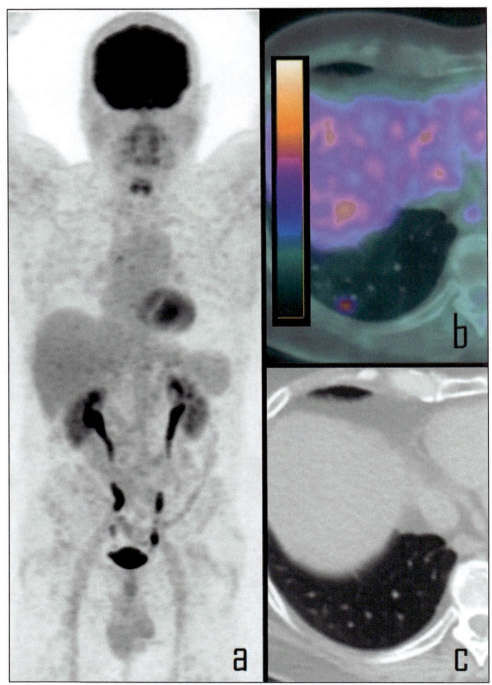

FIGURE 1.7 A 63-year-old man examined during follow-up of colon cancer, two years after surgery. Whole body ^{18}F-FDG PET (a) did not show areas of pathologic tracer uptake. Nevertheless, a focus of increased uptake was detected in the posterior basal segment of the right lung, as evident in fused PET/CT (b). This finding was not associated to meaningful morphological abnormalities at CT imaging (c). This finding was considered as false positive. *Sine materia* uptake may be linked to vascular phase of injected tracer [22].

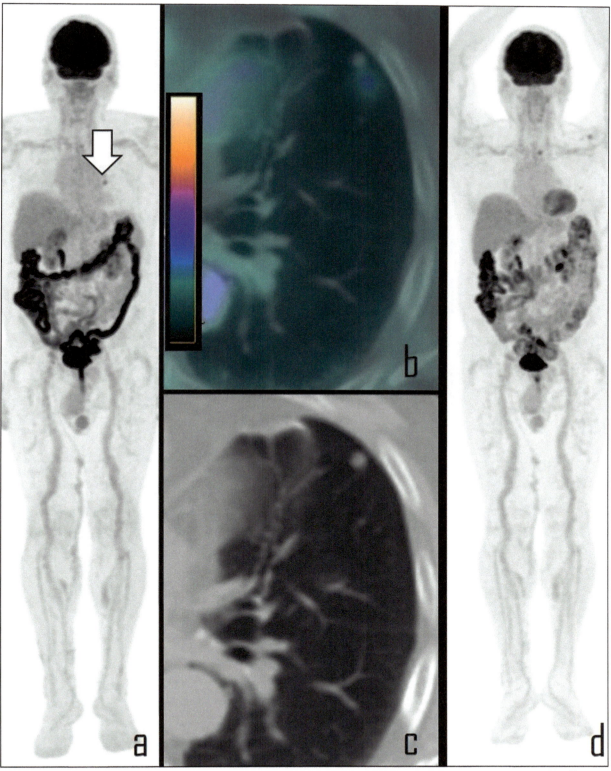

FIGURE 1.8 A 54-year-old man examined during follow-up, 3 years after surgical excision of dorsal cutaneous melanoma. Whole body ^{18}F-FDG PET (a, *arrow*) showed focal, mild uptake in the left lung. Axial PET/CT (b) confirmed mild uptake in a solitary, 4-mm-wide lung node, despite motion respiratory artifacts in fused imaging, with SUVmax 2.3. Despite low SUVmax, correlative CT allowed suspicion of melanoma metastasis, due to absence of this finding in a previous PET/CT scan, performed one year before (d). Small pulmonary malignant lesions may show mild tracer uptake, due to tiny size and PET power resolution limit rather than biological behavior.

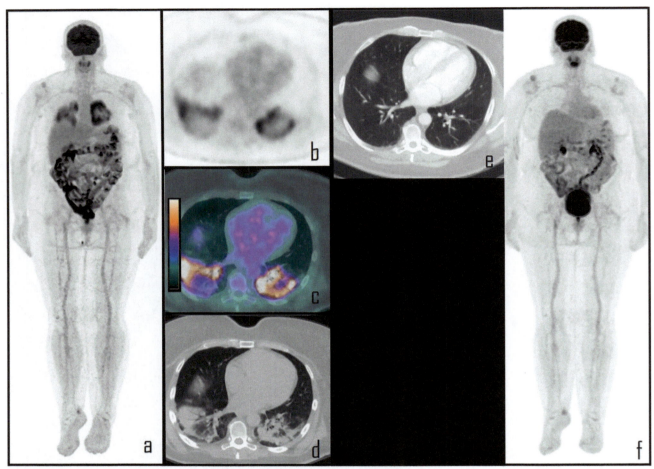

FIGURE 1.9 A 61-year-old patient was submitted to surgical excision of cutaneous melanoma of the neck. One year after, this patient underwent ^{18}F-FDG PET/CT as follow-up. Whole body PET (a) showed high tracer uptake in both lung, as also evident in axial PET (b) and PET/CT (c) views. Correlative CT showed bilateral pulmonary pneumonitis, with air bronchogram. In this false-positive case, the registered high SUVmax (9.8) did not correlate with a malignant condition. Axial detail of contrast-enhanced CT performed after antibiotic therapy shows resolution of pneumonitis (e), as well as whole body PET (f), performed during follow-up, one year later.

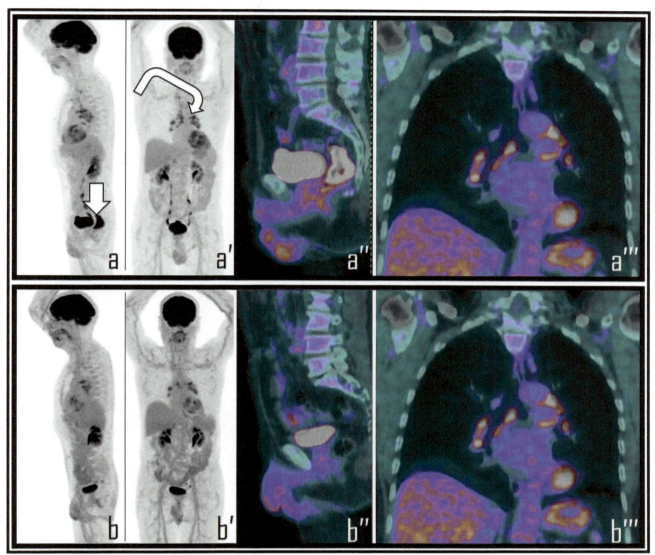

FIGURE 1.10 A 47-year-old patient with known history of thoracic sarcoidosis undergone ¹⁸F-FDG PET/CT during staging of rectal cancer. Baseline ¹⁸F-FDG PET/CT *Maximum Intensity Projection* in sagittal (a) and coronal (a') views shows pathologic tracer uptake in primary rectal lesion (*arrow*) and intense uptake in mediastinal lymph nodes (*curved arrow*), due to concurrent active sarcoidosis. These findings are better shown, respectively, in sagittal PET/CT view of the pelvis (a'') and coronal PET/CT view of the thorax (a'''). A further PET/CT scan, performed after adjuvant chemotherapy, shows reduction of the uptake in rectal cancer and persistence of the uptake in mediastinal lymph nodes, due to sarcoidosis. These results are summarized in sagittal (b) and coronal (b') ¹⁸F-FDG PET/CT *Maximum Intensity Projection* views, sagittal PET/CT detail of the pelvis (b'') and coronal PET/CT view of the thorax (b'''). Sarcoidosis may be a topic of interest, being a benign disease which is adequately depicted by means of ¹⁸F-FDG PET/CT. In particular, this diagnostic tool can help in clinical diagnosis of active sarcoidosis and can play a role in assessment of response to therapy [23]. Sarcoidosis may also be a source of false positive cases in oncologic patients.

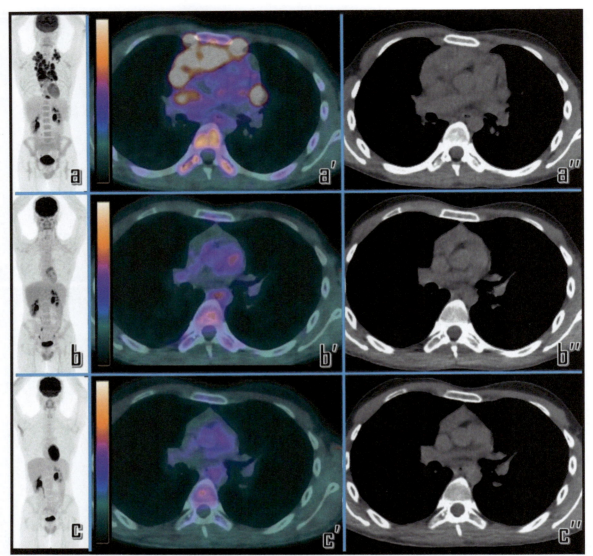

FIGURE 1.11 A 26-year-old man was examined by means of ^{18}F-FDG PET/CT for staging Hodgkin's lymphoma. Several supradiaphragmatic sites of pathologic tracer uptake were detected in whole body PET (a) corresponding to mediastinal pathologic lymph nodes in axial PET/CT (a') and low-dose CT (a'') views. Patient underwent two cycles of chemotherapy, following ABVD regimen (Adriamycin, Bleomycin, Vinblastine, Dacarbazine). Interim PET/CT after initial chemotherapy was negative (b) despite persistence of residual lymphoid tissue, with loss of volume, in the upper mediastinum in PET/CT (b') and CT (c'') views. The last PET/CT, performed at the end of chemotherapy protocol, was negative (c), with further reduction in size of the residual tissue in the upper mediastinum. Interim PET/CT is a fundamental diagnostic tool in management of Hodgkin's lymphoma patients. In fact, it shows early response to chemotherapy allowing, in a minority of nonresponders, the possibility to switch to further second-line protocols. Moreover, a negative interim ^{18}F-FDG PET/CT may also predict final response to therapy and overall survival with very low rate progression-free survival in treated individuals with Hodgkin's lymphoma [24]. The accurate sensitivity of PET can overcome limitations of detectable lymphadenopathies which may last detectable at CT for several months also after the end of the therapy.

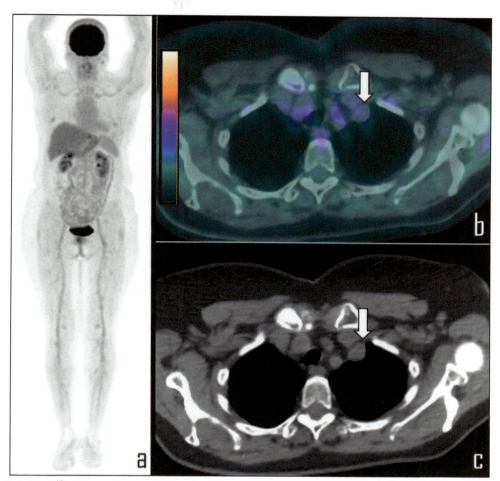

FIGURE 1.12 Whole body ^{18}F-FDG PET *Maximum Intensity Projection* (a) of a 72-year-old woman, not showing pathological tracer uptake. The patient was submitted, 2 years prior to the exam, to surgical excision of liposarcoma of the left leg. Axial PET/CT (b) and CT (c) views of the thorax show a 2-cm-wide hypodense node, not ^{18}F-FDG-avid, in the anterior segment of the left superior pulmonary lobe (*arrow*), due to liposarcoma recurrence. Malignant lipomatous tumors as liposarcoma can show faint tracer uptake [25]. The accurate analysis of coregistered CT of PET/CT is of the utmost importance to avoid misdiagnosis.

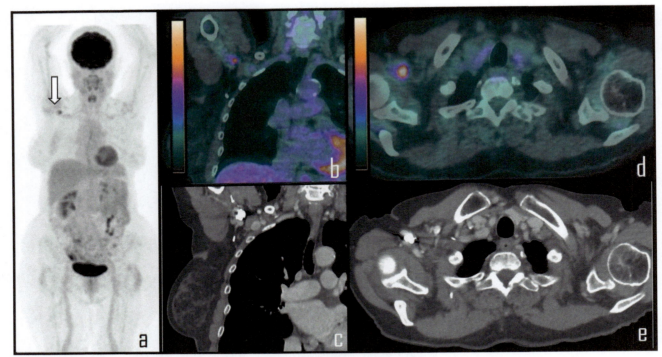

FIGURE 1.13 Whole body ^{18}F-FDG PET *Maximum Intensity Projection* (a) of a 65-year-old woman, during follow-up of mammarian cancer. No pathological tracer uptake was detected. A focal area of ^{18}F-FDG uptake can be observed in right axilla in coronal PET/CT view (b) and correlative coronal contrast-enhanced CT view (c). The uptake is due to intravenous stasis, also evident with iodinated contrast agent. This finding is also shown in axial PET/CT (d) and contrast-enhanced (e) views.

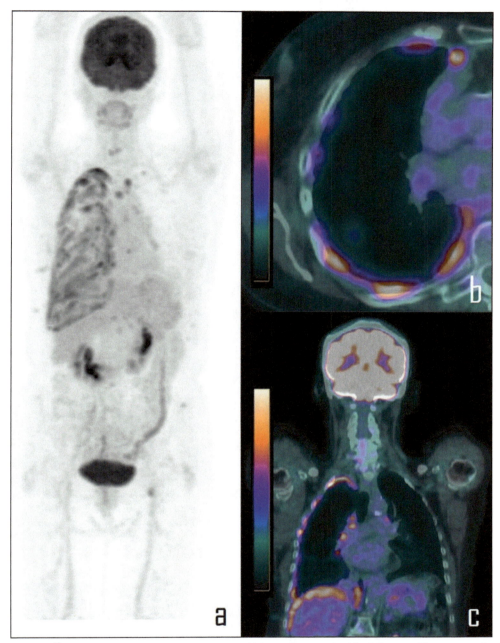

FIGURE 1.14 Whole body ^{18}F-FDG PET *Maximum Intensity Projection* (a) in a patient examined for follow-up of lung cancer, showing multiple areas of intense uptake in right hemithorax, corresponding to extensive nodular thickening of the right pleura due to pleural carcinomatosis, as showed in axial PET/CT detail (b) and coronal PET/CT view (c).

142 Atlas of Hybrid Imaging

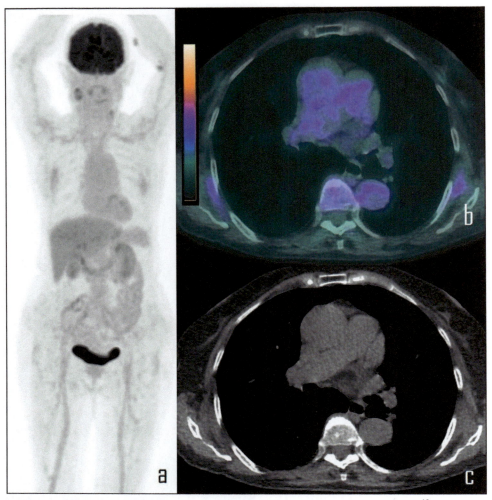

FIGURE 1.15 A patient examined during follow-up of colon cancer, five years following left hemicolectomy. ^{18}F-FDG PET *Maximum Intensity Projection* (a) shows mild tracer uptake in the thorax, bilaterally, corresponding to soft tissue infrascapular masses, showed in axial PET/CT (b) and CT (c) views, recognizable as *elastofibroma dorsi*.

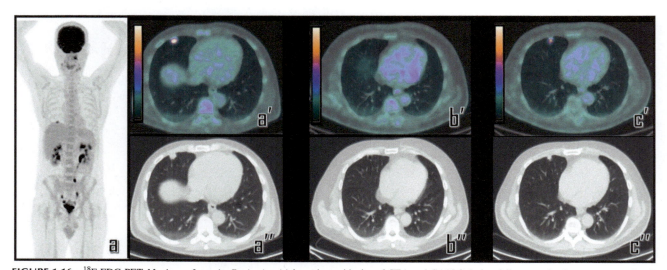

FIGURE 1.16 ^{18}F-FDG PET *Maximum Intensity Projection* (a) in patient with rise of CEA and CA19.9 during follow-up of colon cancer, previously treated with right hemicolectomy, showing focal uptake in the right hemithorax, corresponding to solitary lung metastases, 14 mm wide, with SUVmax 5.9, irregular margins and pleural contact, displayed in correlative axial PET/CT (a') and CT (a'') views. Patient underwent chemotherapy. A further PET/CT scan demonstrated response to therapy, with area of cavitation without metabolic activity (b', b''). However, a third exam showed tumor relapse with SUVmax 4.9 in the same lesion (c', c'').

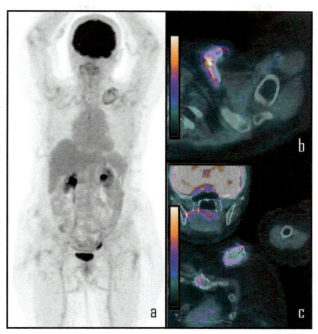

FIGURE 1.17 A patient submitted to implant of pacemaker in left pectoral region presented persistent fever without response to antibiotic therapy, one year after implantation. ^{18}F-FDG PET *Maximum Intensity Projection* (a) showed increased tracer uptake in left pectoral region, close to the pacing box, as evident in axial (b) and coronal (c) PET/CT views, indicating pocket infection.

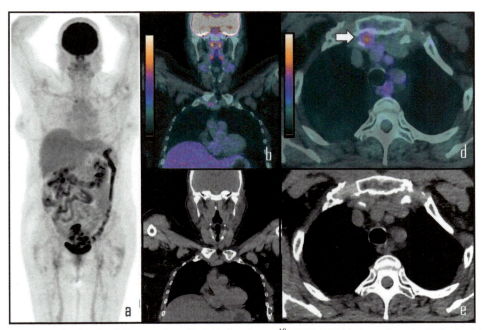

FIGURE 1.18 In a patient examined during follow-up of non-Hodgkin lymphoma, ^{18}F-FDG PET *Maximum Intensity Projection* (a) shows focal uptake in the upper thorax, corresponding to soft tissues close to the right sternoclavicular joint, as displayed in coronal PET/CT (b) and CT (c) views and correlative axial PET/CT (d) and CT (e) views. Markedly increased ^{18}F-FDG uptake involving the right sternoclavicular joint was due to arthritic degeneration, as detectable in CT images.

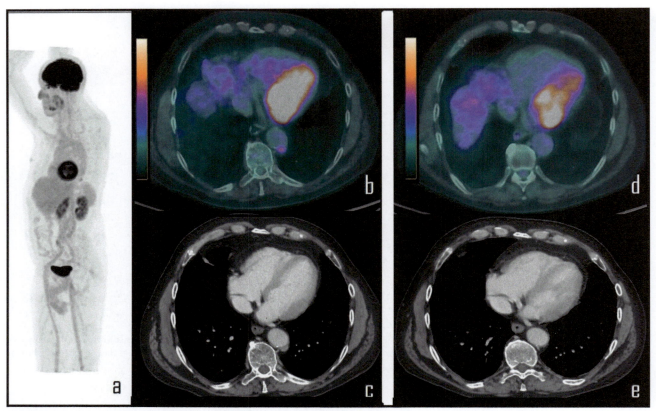

FIGURE 1.19 A 37-year-old patient examined by means of ^{18}F-FDG PET/CT during follow-up of seminoma, one year after right orchifunicolectomy. ^{18}F-FDG PET *Maximum Intensity Projection* (a) in anterior left oblique three-dimensional view shows focal uptake in the inferior mediastinum, corresponding to a small, 7-mm-wide metastatic lymph node in PET/CT view (b). This finding was not clearly pathologic at contrast-enhanced CT, performed in consensus with PET/CT, as evident in correlative axial CT view (c). Patient underwent radiotherapy. A further contrast-enhanced PET/CT exam showed complete response to therapy, as displayed in axial PET/CT (d) and contrast-enhanced CT (e) views. One of the major advantages of PET/CT is the capability of an early diagnosis of tumor relapse, in comparison with CT and MRI, allowing identification of oligometastatic patients and improving proper treatment and overall survival [26].

1.3.2 ^{18}F-DOPA

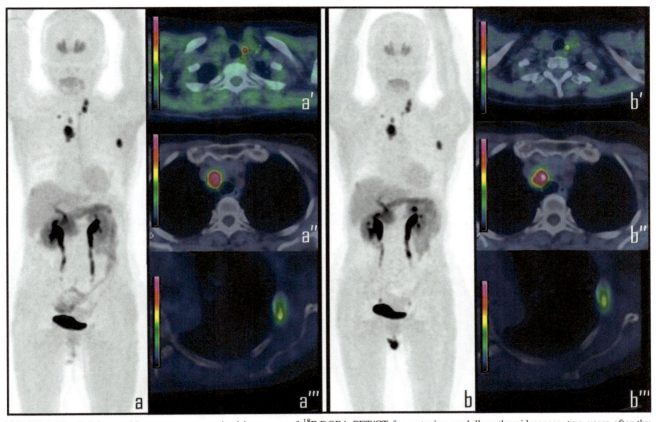

FIGURE 1.20 A 36-year-old woman was examined by means of ^{18}F-DOPA PET/CT for restaging medullary thyroid cancer, two years after thyroidectomy. Whole body ^{18}F-DOPA PET (a) showed several area of pathologic tracer uptake, corresponding to local relapse in thyroid loggia (a'), lymph nodal mediastinal metastasis, 2.5 cm wide, in the Barety's space (a'') and osteolytic lesion with bulging of the cortex in a rib of the left hemithorax (a'''), as evident in cited PET/CT views. Low-dose CT, with adequate window level, is usually sufficient to adequately identify and depict sites of abnormal tracer uptake at PET imaging [27]. Patient refused therapy; interestingly, no progression disease was observed at ^{18}F-DOPA PET/CT, performed at follow-up (b, b', b'', b'''), one year after. Patients with thyroid medullary cancer may be eloquently evaluated with ^{18}F-DOPA PET/CT [28].

1.3.3 ^{68}Ga-DOTATOC

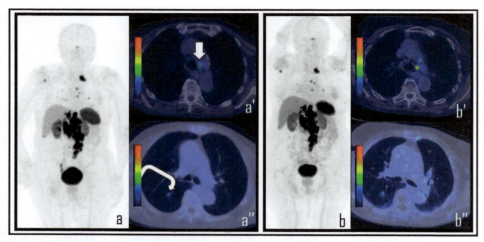

FIGURE 1.21 A 55-year-old female patient, previously submitted to surgical excision of neuroendocrine tumor of the ileum, was examined by means of ^{68}Ga-DOTATOC PET/CT. Whole body ^{68}Ga-DOTATOC PET *Maximum Intensity Projection* (a) demonstrated multiple lymph node metastases with high expression of somatostatine receptors in the abdomen, also showing supra-diaphragmatic lesions, corresponding to bony localizations in the cranial teca and right omerus, cervical and mediastinal lymph node localizations (a', *arrow*) and in a small solitary lung node in the upper segment of the inferior right lobe (a'', *curved arrow*), as showed in related PET/CT views. After six months, treatment with somatostatin *cold analogs*, a further ^{68}Ga-DOTATOC PET/CT demonstrated treatment failure (b, b', b'').

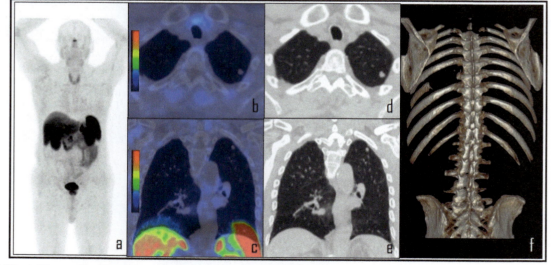

FIGURE 1.22 A 71-year-old male patient was examined with whole body ^{68}Ga-DOTATOC PET/CT for characterization of a solitary lung node associated with rise of *chromogranin A* serum levels (161 ng/mL at the time of the scan), for suspicion of a pulmonary neuroendocrine tumor. No pathologic tracer uptake was detected in whole body PET (a) as well as in the solitary, 1-cm-wide apical node in the left lung, evident in related axial (b) and coronal (c) PET/CT views and in axial (d) and coronal (e) CT views. Volume rendering CT of the skeleton (f) shows absence of the medial portion of the eighth left rib. Patient anamnesis was positive for posttraumatic injury of the left hemithorax, with following partial surgical excision of this rib. Volume rendering of *low-dose* CT of a PET/CT scan is a three-dimensional reconstruction which may improve confidence in detection of skeletal abnormalities, providing most intuitive three-dimensional skeletal images.

1.3.4 ^{18}F-choline

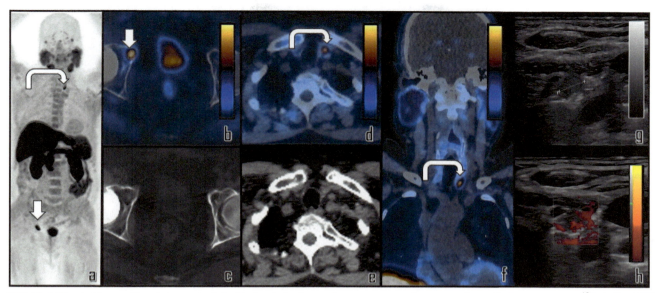

FIGURE 1.23 A 68-year-old patient examined for biochemical relapse of prostate cancer, three years after radical prostatectomy. ^{18}F-choline PET *Maximum Intensity Projection* (a) shows two areas of intense uptake respectively in the pelvis (*arrow*) and in the upper mediastinum (*curved arrow*). Uptake in the pelvis was observed in right ischium, without morphological abnormalities in axial PET/CT (b, *arrow*) and CT (c) views, with bone window level. Considering the PSA at the time of the scan (2.4 ng/mL), this finding was considered as prostate cancer bony lesion. PET can detect bony lesions prior to that anatomical alterations can be observed at CT imaging [29]. The uptake in the upper mediastinum was linked to a subcentimetric hypodense node in axial PET/CT (d, *curved arrow*) and CT (e) views and coronal PET/CT (f, *curved arrow*). Due to suspicion of lesion with high rate lipogenesis, other than prostate cancer, patient underwent ultrasonography of the neck, showing 7-mm-wide, hypoechogenic node (g) with vascular spots at power Doppler. Hematic parathyroid hormone resulted 170 pg/mL. Final diagnosis was parathyroid adenoma [30].

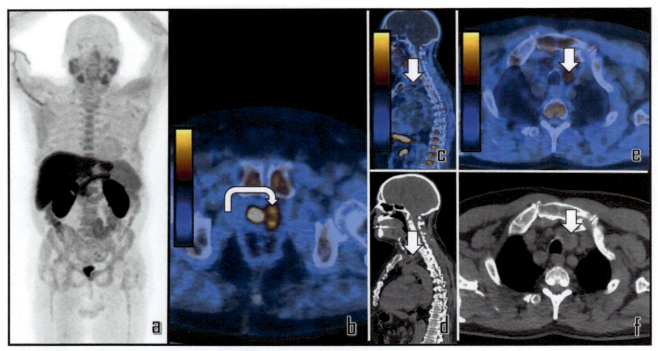

FIGURE 1.24 A 71-year-old patient was examined by ^{18}F-choline PET/CT for biochemical recurrence of prostate cancer. Laboratory data expressed rapid PSA kinetics (PSA 2.1 ng/mL; PSA velocity 0.3 ng/mL/year; PSA doubling time 5 months). The patient underwent one year before radical prostatectomy. Whole body PET (a) showed focal uptake in the prostate bed, more evident in axial PET/CT view (b, *curved arrow*): this finding was considered local relapse. PSA at the time of the scan correlates with a positive ^{18}F-choline PET/CT. Moreover, PSA kinetics as *PSA doubling* time and *PSA velocity* positively affects the sensitivity of the exam, also improving confidence in diagnosis [31]. As secondary finding, mild tracer uptake was observed in an intrathoracic mediastinal thyroid goiter, evident in sagittal PET/CT (c, *arrow*) and CT (d, *arrow*) views and axial PET/CT (e, *arrow*) and CT (f, *arrow*) views. Thyroid uptake of radiolabeled choline is common and due to functional processes or inflammation [32]. Disomogeneal hypodensity in the upper mediastinum, linked to thyroid parenchyma, may suggest diagnosis.

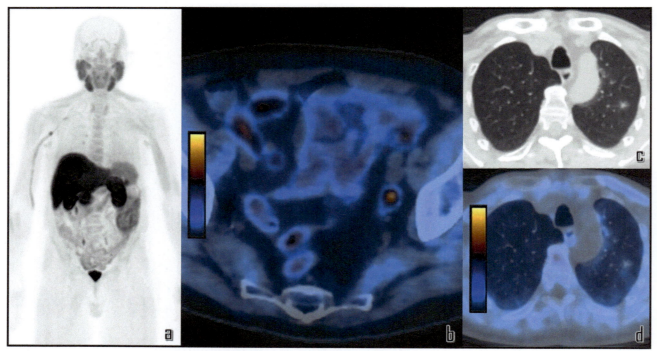

FIGURE 1.25 A patient with prostate cancer, previously submitted to radical prostatectomy, underwent ^{18}F-choline PET/CT due to biochemical relapse (PSA 0.9 ng/mL). Whole body PET *Maximum Intensity Projection* (a) shows single area of focal uptake in the left pelvis, corresponding to a solitary, 8-mm-wide, left iliac lymph node, considered as prostate cancer relapse. The main advantage of ^{18}F-choline PET/CT is linked to the possibility to diagnose lymph node metastases also in small lymph nodes, not clearly pathologic in contrast-enhanced CT [29] providing earlier diagnosis and improving targeted treatment and patients' survival. In the same patient, mild uptake was documented in a peripheral hyperdense node with poorly outlined edges, in the lower left lobe, as evident in axial CT (c) and PET/CT (d) views with lung CT window level. This finding was considered as inflammation [33].

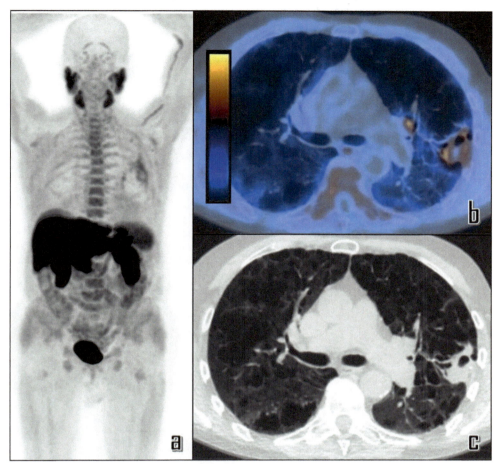

FIGURE 1.26 A patient with prostate cancer, previously submitted to radiotherapy, underwent ^{18}F-choline PET/CT (PSA 1.2 ng/mL at the time of the scan). Whole body PET (a) was negative. However, biochemical relapse after radiotherapy can be considered when PSA serum level is higher than 1.0 ng/mL, due to a physiologic production of this marker in residual prostate gland [34]. In this patient, mild and diffuse uptake was documented in an area of parenchymal consolidation in left lung, with internal cavitation. Further area of uptake was recorded in centimetric lymph node in left hilar region. Patient's anamnesis was positive for productive cough and low-grade fever in the month preceding the scan. These findings, summarized in axial PET/CT (b) and CT (c) views were considered as pneumonitis. Inflammatory processes are at the basis of the large majority of false-positive cases in PET/CT with ^{18}F-choline. Generally, the diagnosis of inflammation is easily recognizable with accurate anamnesis [35].

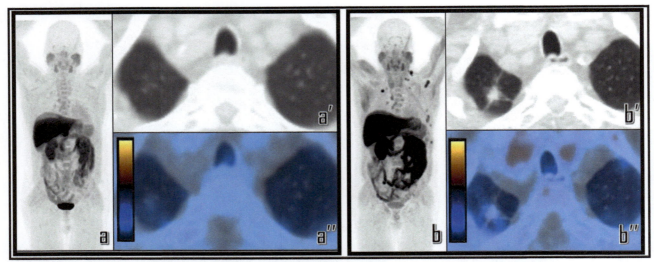

FIGURE 1.27 A 63-year-old patient undergone ^{18}F-choline PET/CT for early biochemical relapse of prostate cancer (PSA 0.2 ng/mL), 9 months after radical prostatectomy. Whole body ^{18}F-choline PET was negative; no pulmonary lesions were detected in PET/CT (a') and CT (a''). One year after, PSA was considerably higher (8.2 ng/mL) and the patient underwent a further PET/CT scan, showing several bony localizations of prostate cancer (b). Interestingly, an area of parenchymal consolidation was detected in the apex of the right lung, not ^{18}F-choline-avid. This finding, not present at previous PET/CT, was examined by biopsy: histological exam diagnosed spinocellular carcinoma.

1.3.5 ^{18}F-NaF

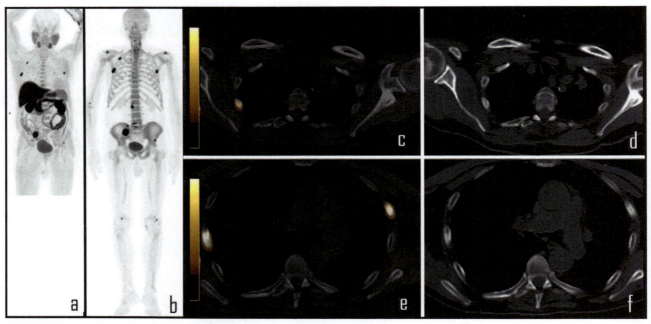

FIGURE 1.28 A patient in biochemical relapse of prostate cancer (PSA 6.9 ng/mL) was examined with ^{18}F-choline PET/CT (a), showing four bony lesions in the thorax and in the pelvis. Whole body ^{18}F-NaF PET (b) confirmed these findings, correctly identifying more lesions in the column, in the right omerus and left clavicle, due to better sensitivity of ^{18}F-NaF PET/CT in identification bone metastases, in comparison to radiolabeled choline and other PET radiopharmaceuticals [36]. Axial ^{18}F-NaF PET/CT view of the thorax (c) shows a tracer-avid lesion in the third right rib, without corresponding morphological abnormalities in correlative CT view (d). Bone metastasis detection with ^{18}F-NaF PET/CT may be considerably precocious in comparison to morphological changes detectable in CT [37]. On the other hand, another ^{18}F-NaF axial PET/CT view (e) displays further bilateral rib metastases, in association with sclerotic lesions on CT imaging (f). The CT bone window is of the utmost importance for the optimal evaluation of bone structures of the thorax as well as of the entire body.

1.3.6 ^{131}I

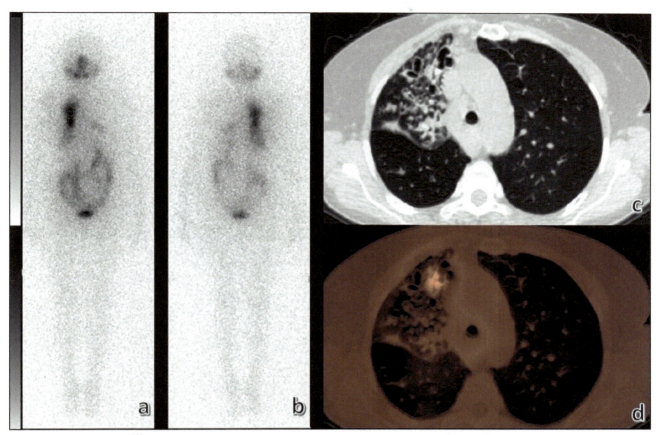

FIGURE 1.29 Postsurgical ^{131}I whole body scan in a patient with thyroid cancer (a, b) shows tracer uptake in the right hemithorax, in correspondence of bronchiectasis of the middle lobe of the right lung, in correlative axial CT (c) and SPECT/CT (d) views.

152 Atlas of Hybrid Imaging

1.3.7 ^{99m}Tc-MDP

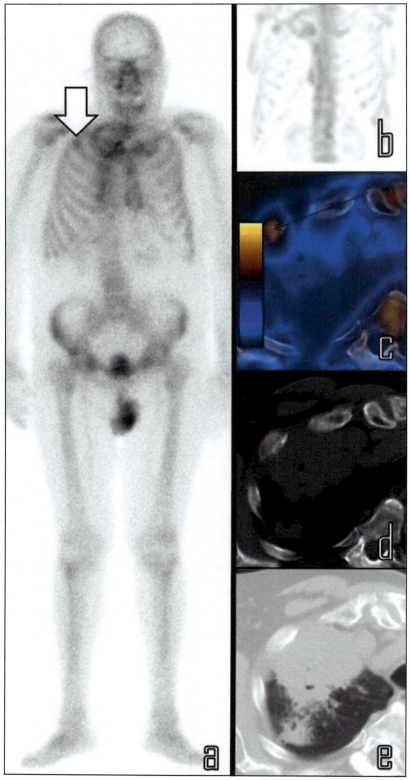

FIGURE 1.30 During the staging of lung cancer, a 73-year-old man was examined with ^{99m}Tc-MDP bone scan, showing single focus of increased tracer uptake in the second rib of the right hemithorax (a, *arrow*), subsequently confirmed at SPECT (b) and SPECT/CT imaging (c, *arrow*). Correlative CT with bone window level (d) showed this finding was associated to a millimetric lytic lesion of the rib, while the same axial view with lung window (e) showed close proximity of the rib to primitive pulmonary tumor. Final diagnosis was single bone metastasis of the rib due to contiguity to primitive lung lesion. The adequate window level of low-dose CT of nuclear imaging scans can help in correctly identifying pathologic foci of uptake, allowing the adequate depiction of associated structural changes and anatomical abnormalities. Specificity of ^{99m}Tc-MDP bone scan may be significantly improved by hybrid SPECT/CT imaging, when required [38].

Thorax Chapter | 1   153

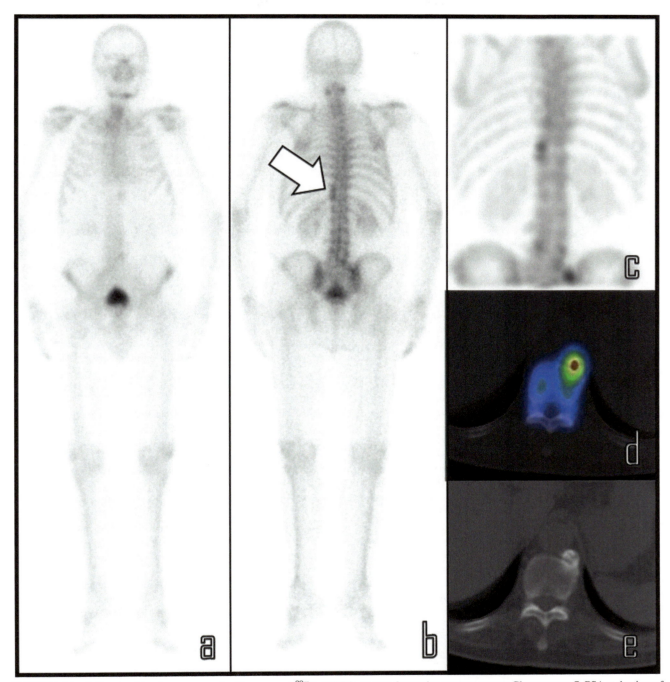

FIGURE 1.31 A 61-year-old patient was examined by means of ^{99m}Tc-MDP bone scan for staging prostate cancer; Gleason score 7. PSA at the time of the scan was 8.2 ng/mL. Bone scan in anterior (a) and posterior (b, *arrow*) views shows a single area of focal uptake in the distal tract of the dorsal column, confirmed by SPECT (c). Axial SPECT/CT view (d) displays this finding as linked to osteophitosis of the 10th thoracic vertebra, more evident in related CT view with bone window level (e).

References

[1] Timmerman R, McGarry R, Yiannoutsos C, Papiez L, Tudor K, DeLuca J, et al. Excessive toxicity when treating central tumors in a phase II study of stereotactic body radiation therapy for medically inoperable early-stage lung cancer. J Clin Oncol 2006;24:4833−9.

[2] Proto AV, JBall Jr JB. Computed tomography of the major and minor fissures. AJR Am J Roentgenol 1983;140:439−48.

[3] Osborne D, Vock P, Godwin JD, Silverman PM. CT identification of bronchopulmonary segments: 50 normal subjects. AJR Am J Roentgenol 1984;142:47−52.

[4] Chiumello D, Marino A, Brioni M, Menga F, Cigada I, Lazzerini M, et al. Visual anatomical lung CT scan assessment of lung recruitability. Intensive Care Med 2013;39:66−73.

[5] Goerres GW, Kamel E, Seifert B, Burger C, Buck A, Hany TF, et al. Accuracy of image coregistration of pulmonary lesions in patients with non-small cell lung cancer using an integrated PET/CT system. J Nucl Med 2002;43:1469−75.

[6] Grootjans W, Hermsen R, van der Heijden EHFM, Schuurbiers-Siebers OCJ, Visser EP, Oyen WJG, et al. The impact of respiratory gated positron emission tomography on clinical staging and management of patients with lung cancer. Lung Cancer 2015;90:217−23.

[7] Rinaldi R, Camoni L, Albano D. Performing an additional lateral decubitus PET/CT scan to resolve a respiratory motion artifact. J Nucl Med Technol 2020;49:84−5. Online ahead of print.

[8] Fukushima N, Shimojima N, Ishitate M. Clinical and structural aspects of tracheal stenosis and a novel embryological hypothesis of left pulmonary artery sling. Pediatr Pulmonol 2020;55:747−53.

[9] Williams PL, Warwick R, Dyson M, Bannister LH. Splanchnology. In: Gray's anatomy. 37th ed. New York, NY: Churchill Livingstone; 1989. p. 1245−475.

[10] Carter BW, Tomiyama N, Bhora FY, Rosado de Christenson ML, Nakajima J, Boiselle PM, et al. A modern definition of mediastinal compartments. J Thorac Oncol 2014;9:S97−101.

[11] Mallorie A, Goldring J, Patel A, Lim E, Wagner T. Assessment of nodal involvement in non-small-cell lung cancer with 18F-FDG-PET/CT: mediastinal blood pool cut-off has the highest sensitivity and tumour SUVmax/2 has the highest specificity. Nucl Med Commun 2017;38:715−9.

[12] Gallamini A, Barrington SF, Biggi A, Chauvie S, Kostakoglu L, Gregianin M, et al. The predictive role of interim positron emission tomography for Hodgkin lymphoma treatment outcome is confirmed using the interpretation criteria of the Deauville five-point scale. Haematologica 2014;99:1107−13.

[13] Sagheb S, Metser U, Razaz S, Menezes R, Gallinger S, Jhaveri KS. Preliminary evaluation of 18F-FDG-PET/MRI for differentiation of serous from nonserous pancreatic cystic neoplasms: a pilot study. Nucl Med Commun 2020;41:1257−64.

[14] Zincirkeser S, Sahin E, Halac M, Sager S. Standardized uptake values of normal organs on 18F-fluorodeoxyglucose positron emission tomography and computed tomography imaging. J Int Med Res 2007;35:231−6.

[15] Nakamoto R, Okuyama C, Ishizu K, Higashi T, Takahashi M, Kusano K, et al. Diffusely decreased liver uptake on FDG PET and cancer-associated cachexia with reduced survival. Clin Nucl Med 2019;44:634−42.

[16] Gorospe L, Raman S, Echeveste J, Avril N, Herrero Y, Herna Ndez S. Whole-body PET/CT: spectrum of physiological variants, artifacts and interpretative pitfalls in cancer patients. Nucl Med Commun 2005;26:671−87.

[17] Priola AM, Priola SM. Chemical-shift MRI of rebound thymic hyperplasia with unusual appearance and intense (18)F-FDG uptake in adulthood: report of two cases. Clin Imaging 2014;38:739−42.

[18] Al-Mnayyis A, Al-Alami Z, Altamimi N, Alawneh KZ, Aleshawi A. Azygos lobe: prevalence of an anatomical variant and its recognition among postgraduate physicians. Diagnostics (Basel) 2020;10:470.

[19] Schillaci O, Travascio L, Bolacchi F, Calabria F, Bruni C, Cicciò C, et al. Accuracy of early and delayed FDG PET-CT and of contrast-enhanced CT in the evaluation of lung nodules: a preliminary study on 30 patients. Radiol Med 2009;114:890−906.

[20] Subramanian DR, Jenkins L, Edgar R, Quraishi N, Stockley RA, Parr DG. Assessment of pulmonary neutrophilic inflammation in emphysema by quantitative positron emission tomography. Am J Respir Crit Care Med 2012;186:1125−32.

[21] Ganem J, Thureau S, Gardin I, Modzelewski R, Hapdey S, Vera P. Delineation of lung cancer with FDG PET/CT during radiation therapy. Radiat Oncol 2018;13:219.

[22] Chondrogiannis S, Marzola MC, Grassetto G, Zorzi A, Milan E, Rampin L, et al. 18F-FDG PET/CT lung 'focalities' without coregistered CT findings: an interpretative clinical dilemma. Nucl Med Commun 2015;36:334−9.

[23] Tetikkurt C, Yanardag H, Sayman BH, Bilir M, Tetikkurt S, Bilgic S, et al. Diagnostic utility of 68Ga-citrate and 18FDG PET/CT in sarcoidosis patients. Monaldi Arch Chest Dis 2020;90:4.

[24] Gallamini A, Zwarthoed C. Interim FDG-PET Imaging in Lymphoma. Semin Nucl Med 2018;48:17−27.

[25] Baffour FI, Wenger DE, Broski SM. (18)F-FDG PET/CT imaging features of lipomatous tumors. Am J Nucl Med Mol Imaging 2020;10:74−82.

[26] Rodríguez-Fraile M, Cózar-Santiago MP, Sabaté-Llobera A, Caresia-Aróztegui AP, Delgado Bolton RC, Orcajo-Rincon J, et al. FDG PET/CT in colorectal cancer. Rev Esp Med Nucl Imagen Mol 2020;39:57−66.

[27] Kluijfhout WP, Paternak JD, Drake FT, Beninato T, Gosnell JE, Shen WT. Use of PET tracers for parathyroid localization: a systematic review and meta-analysis. Langenbecks Arch Surg 2016;401:925−35.

[28] Treglia G, Castaldi P, Villani MF, Perotti G, Filice A, Ambrosini V, et al. Comparison of different positron emission tomography tracers in patients with recurrent medullary thyroid carcinoma: our experience and a review of the literature. Recent Results Cancer Res 2013;194:385−93.

[29] Schillaci O, Calabria F, Tavolozza M, Caracciolo CR, Finazzi Agrò E, Miano R, et al. Influence of PSA, PSA velocity and PSA doubling time on contrast-enhanced 18F-choline PET/CT detection rate in patients with rising PSA after radical prostatectomy. Eur J Nucl Med Mol Imaging 2012;39:589−96.

[30] Boccalatte LA, Gómez NL, Musumeci M, Galich AM, Collaud C, Figari MF. (18)F-choline PET/4D CT in hyperparathyroidism: correlation between biochemical data and study parameters. Rev Esp Med Nucl Imagen Mol 2020;39:273—8.

[31] Calabria F, Rubello D, Schillaci O. The optimal timing to perform 18F/11C-choline PET/CT in patients with suspicion of relapse of prostate cancer: trigger PSA versus PSA velocity and PSA doubling time. Int J Biol Markers 2014;29:e423—30.

[32] Calabria F, Chiaravalloti A, Cicciò C, Gangemi V, Gullà D, Rocca F, et al. PET/CT with (18)F-choline: Physiological whole bio-distribution in male and female subjects and diagnostic pitfalls on 1000 prostate cancer patients: (18)F-choline PET/CT bio-distribution and pitfalls. A southern Italian experience. Nucl Med Biol 2017;51:40—54.

[33] Calabria F, Chiaravalloti A, Schillaci O. (18)F-choline PET/CT pitfalls in image interpretation: an update on 300 examined patients with prostate cancer. Clin Nucl Med 2014;39:122—30.

[34] Caroli P, Colangione SP, De Giorgi U, Ghigi G, Celli M, Scarpi E, et al. (68)Ga-PSMA-11 PET/CT-guided stereotactic body radiation therapy retreatment in prostate cancer patients with PSA failure after salvage radiotherapy. Biomedicines 2020;8:536.

[35] Schillaci O, Calabria F, Tavolozza M, Cicciò C, Carlani M, Caracciolo CR, et al. 18F-choline PET/CT physiological distribution and pitfalls in image interpretation: experience in 80 patients with prostate cancer. Nucl Med Commun 2010;31:39—45.

[36] Lapa P, Saraiva T, Silva R, Marques M, Costa G, Lima JP. Superiority of 18F-FNa PET/CT for detecting bone metastases in comparison with other diagnostic imaging modalities. Acta Med Port 2017;30:53—60.

[37] Gerety EL, Lawrence EM, Wason J, Yan H, Hilborne S, Buscombe J, et al. Prospective study evaluating the relative sensitivity of 18F-NaF PET/CT for detecting skeletal metastases from renal cell carcinoma in comparison to multidetector CT and 99mTc-MDP bone scintigraphy, using an adaptive trial design. Ann Oncol 2015;26:2113—8.

[38] Schillaci O. Hybrid SPECT/CT: a new era for SPECT imaging? Eur J Nucl Med Mol Imaging 2005 May;32(5):521—4.

Chapter 2

Abdomen and pelvis

Abstract

The low-dose CT of nuclear imaging scans offers meaningful information concerning abnormal sites of tracer uptake or pathophysiological findings in abdomen and pelvis; more than usual, when evaluating the abdomen, the knowledge of tracers molecular pathways is essential, to correctly identify pathologic data and avoid misdiagnoses. On the other hand, MRI is the best option in the study of pelvis, due to the peculiar anatomy. This chapter is focused on the low-dose CT in general anatomy of the abdomen, with emphasis on the liver and peritoneum. Moreover, a special section is dedicated to PET/MRI of male and female pelvis. Three-dimensional CT views of the abdomen are also provided. In order to give to the reader a reading key for metabolic imaging, we also present and discuss several PET/CT and SPECT/CT clinical cases, tricks and pitfalls with the following radiopharmaceuticals: ^{18}F-FDG, ^{68}Ga-DOTATOC, ^{18}F-choline, ^{68}Ga-PSMA, and ^{99m}Tc.

Keywords: ^{18}F-choline; ^{18}F-FDG; ^{68}Ga-PSMA; Abdomen; Colon cancer; Liver; Pelvis; PET/CT; PET/MRI; Pitfalls

---

Atlas of Hybrid Imaging. https://doi.org/10.1016/B978-0-443-18733-9.00002-6
Copyright © 2023 Elsevier Inc. All rights reserved.

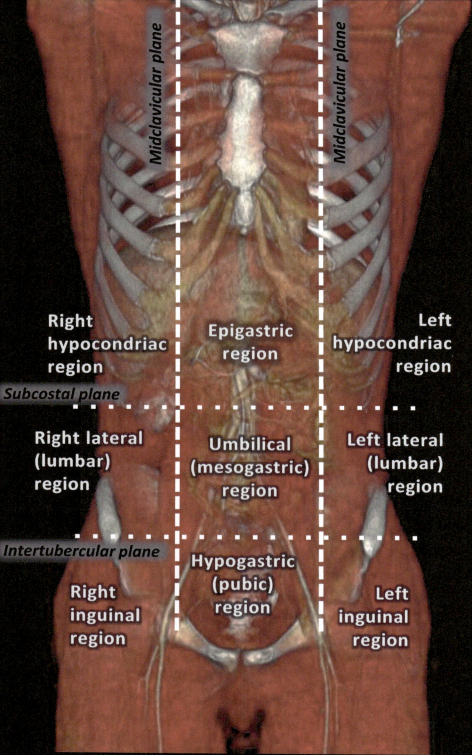

Abdomen and pelvis **Chapter** | **2** **159**

Median plane

Right superior quadrant

Left superior quadrant

Transumbilical plane

Right inferior quadrant

left inferior quadrant

Abdominal surface topography: clinically, the abdominal wall is divided in four quadrants and nine regions, projectively corresponding to subdivisions of the abdominal cavity. This generic subdivision is useful in anatomical localization of tumoral processes especially outside the parenchymatous organs and/or in complex sites at PET/CT.

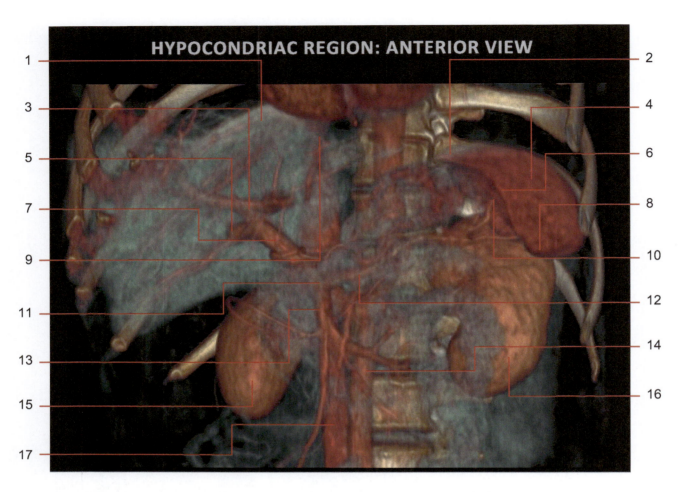

1 Liver
2 Anterior extremity of spleen
3 Left branch of hepatic portal vein
4 Diaphragmatic surface of spleen
5 Right branch of hepatic portal vein
6 Anterior (superior) border of spleen
7 Hepatic portal vein
8 Posterior extremity of spleen
9 Retrohepatic inferior vena cava
10 Splenic hilum and visceral surface of spleen
11 Confluence of superior mesenteric vein and splenic vein
12 Splenic vein
13 Superior mesenteric vein
14 Abdominal aorta
15 Right kidney
16 Left kidney
17 Inferior vena cava

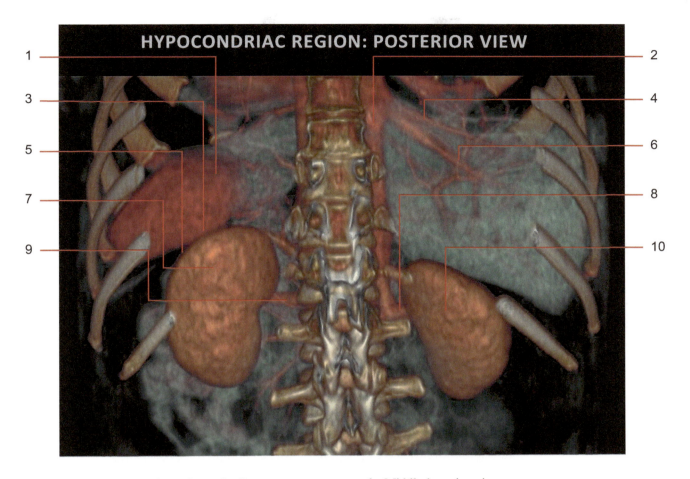

1 Diaphragmatic surface of spleen
2 Retrohepatic inferior vena cava
3 Posterior (inferior) border of spleen
4 Right hepatic vein
5 Splenorenal recess
6 Middle hepatic vein
7 Left kidney
8 Right renal vein
9 Left renal vein
10 Right kidney

162 Atlas of Hybrid Imaging

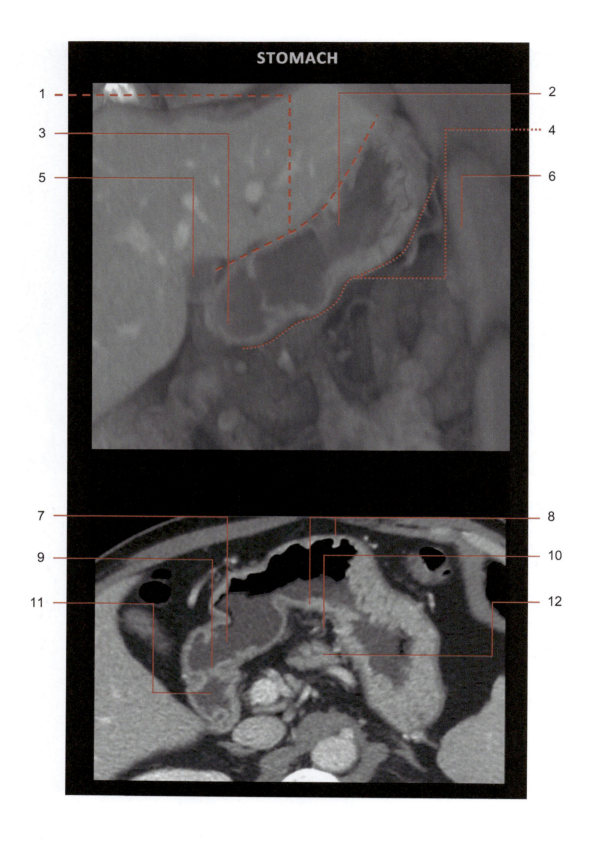

Abdomen and pelvis Chapter | 2

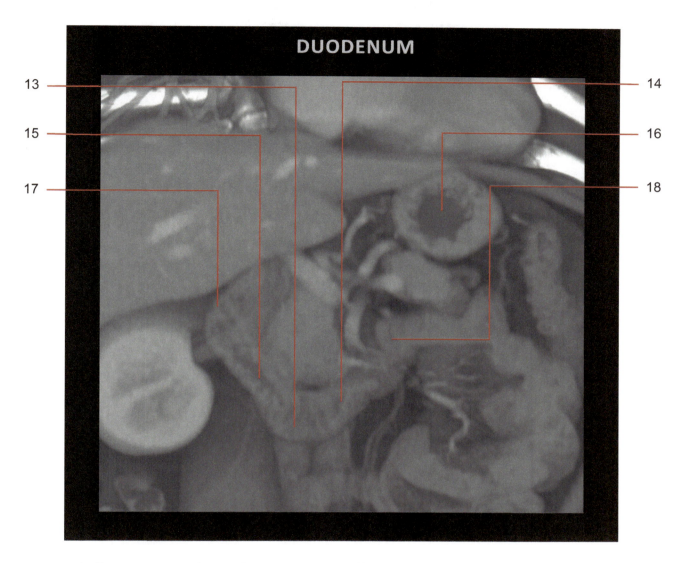

DUODENUM

1 Lesser curvature of stomach
2 Body of stomach
3 Pyloric pars
4 Greater curvature of stomach
5 Gallbladder
6 Spleen
7 Pyloric canal
8 Anterior and posterior walls of stomach
9 Pylorus
10 Omental bursa (lesser sac)
11 Superior (first) part of duodenum (bulb)
12 Pancreas
13 Inferior duodenal flexure
14 Horizontal (third) part of duodenum
15 Descending (second) part of duodenum
16 Fundus of stomach
17 Superior duodenal flexure
18 Ascending (fourth) part of duodenum

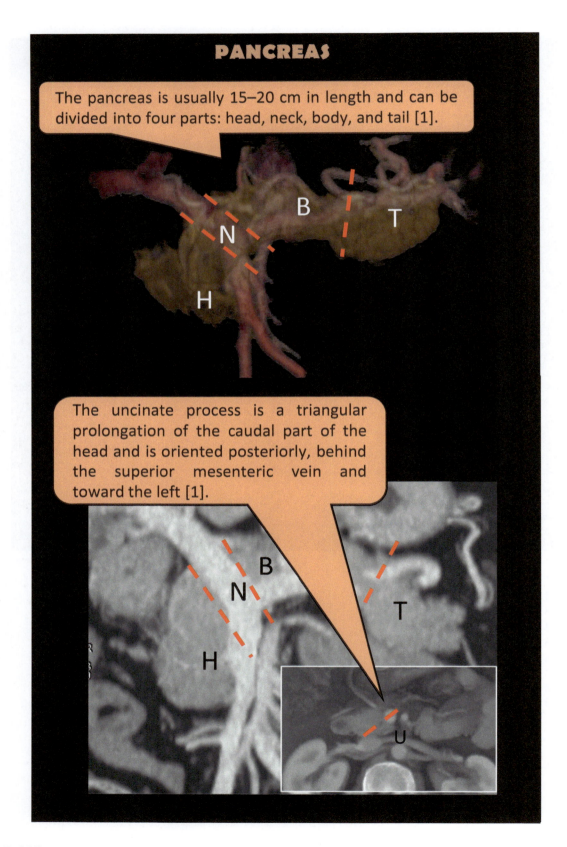

See Ref [1]

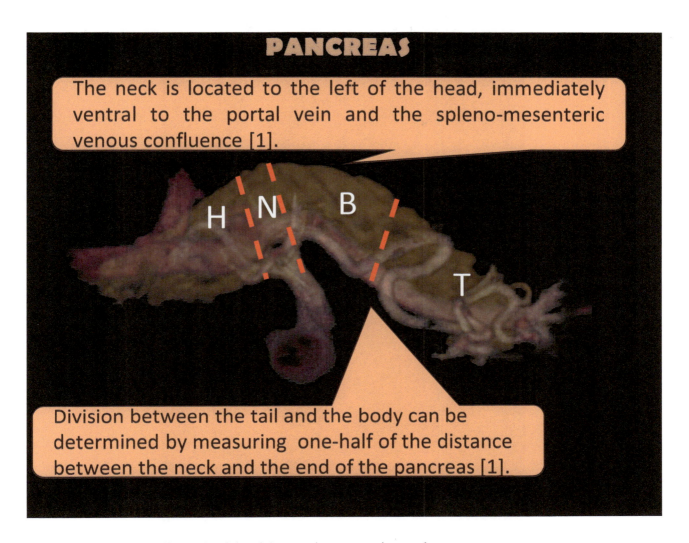

Head (H): lies to the right of the superior mesenteric vessels
Neck (N) or isthmus: lies anterior to superior mesenteric vessels
Body (B): lies to left of superior mesenteric vessels
Tail (T): lies between layers of the splenorenal ligament in the splenic hilum
Uncinate process (U): extension of the head, posterior to superior mesenteric vessels

See Ref [1]

166 Atlas of Hybrid Imaging

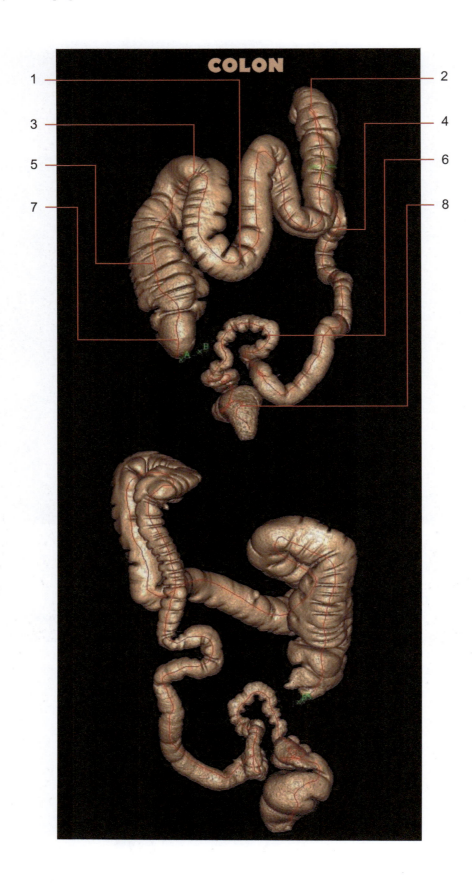

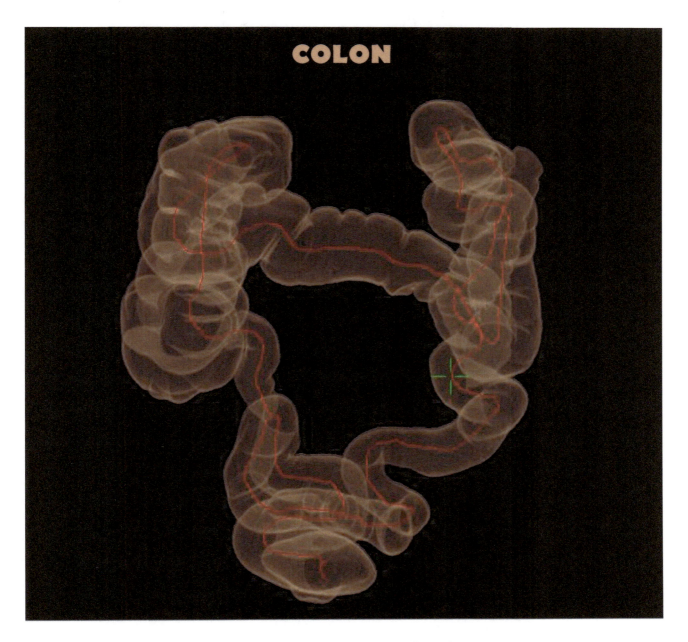

1 Transverse colon	5 Ascending colon
2 Left colic flexure	6 Sigmoid colon
3 Right colic flexure	7 Cecum
4 Descending colon	8 Rectum

See Ref [1]

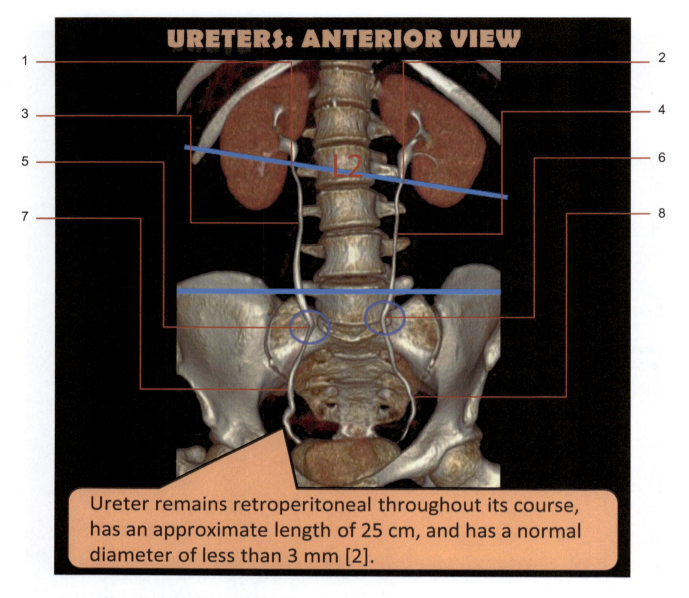

1 Ureteropelvic junction (UPJ) of right ureter
2 Ureteropelvic junction (UPJ) of left ureter
3 Abdominal (lumbar) part of right ureter
4 Abdominal (lumbar) part of left ureter
5 Right ureteral cross (over the iliac vessels)
6 Left ureteral cross (over the iliac vessels)
7 Pelvic part of right ureter
8 Pelvic part of left ureter

See Ref [2]

Abdomen and pelvis Chapter | 2   169

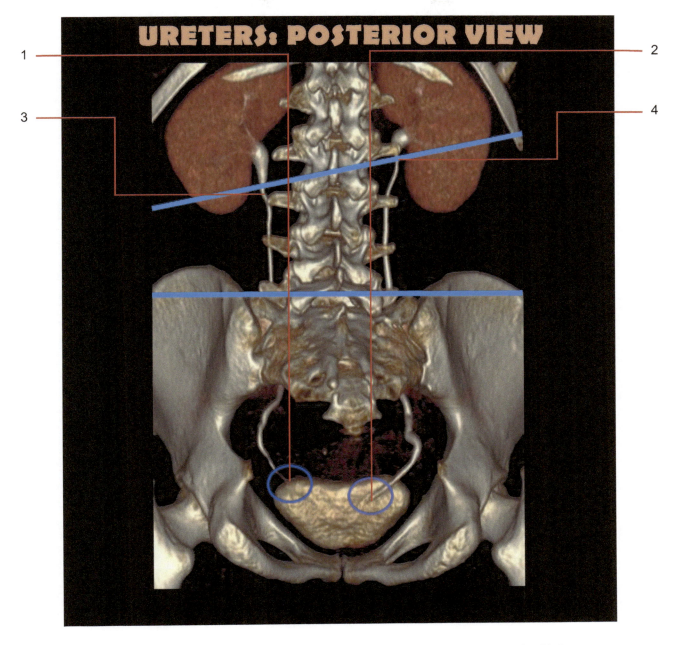

1   Ureterovesicular junction (UVJ) of left ureter
2   Ureterovesicular junction (UVJ) of right ureter
3   Ureteropelvic junction (UPJ) of left ureter
4   Ureteropelvic junction (UPJ) of right ureter

2.1 General anatomy PET/CT
2.1.1 Axial

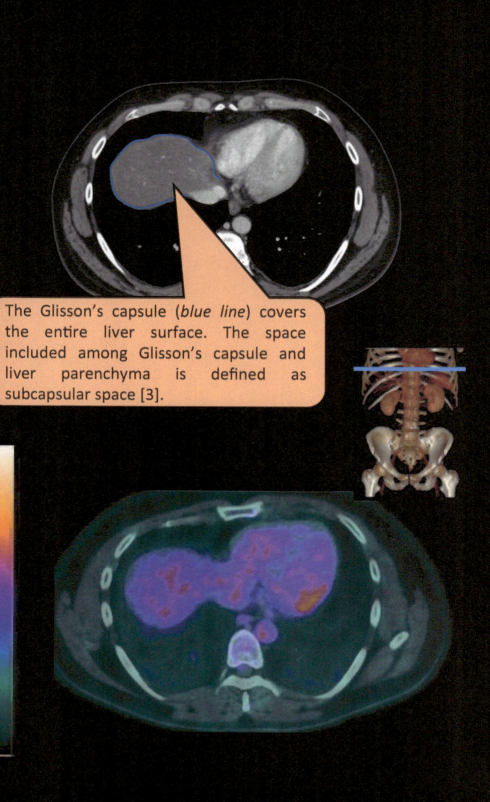

The Glisson's capsule (*blue line*) covers the entire liver surface. The space included among Glisson's capsule and liver parenchyma is defined as subcapsular space [3].

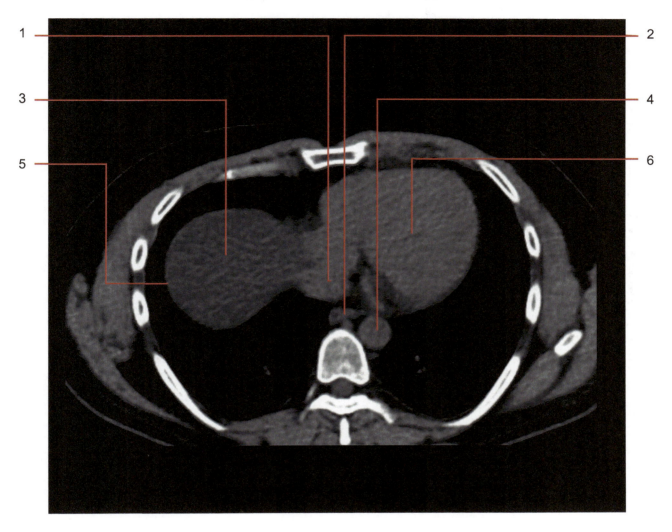

1 Inferior vena cava (suprahepatic)
2 Esophagus
3 Right lobe of liver
4 Thoracic aorta
5 Glisson's capsule (diaphragmatic surface) of liver
6 Heart

CT abdomen (noncontrast) window width (WW) 400
CT abdomen (noncontrast) window level (WL) 40

See Ref [3]

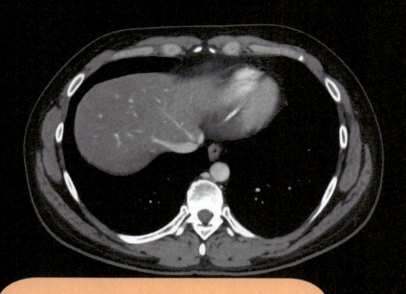

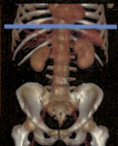

Liver is the standard organ reference for ^{18}F-FDG PET images saturation, in colour or gray scale. Liver uptake is useful for quantification of tumour metabolism [4, 5]. However, the degree of liver uptake may depend from fasting, steatosis and chachexia [6].

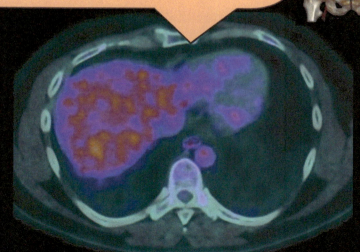

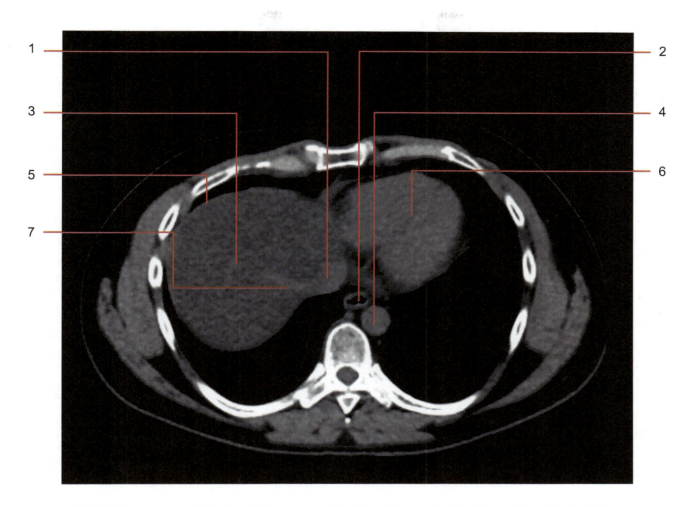

1 Inferior vena cava (retrohepatic)
2 Esophagus
3 Right lobe of liver
4 Thoracic aorta
5 Glisson's capsule (diaphragmatic surface) of liver
6 Heart
7 Hepatic veins

See Refs [4—6]

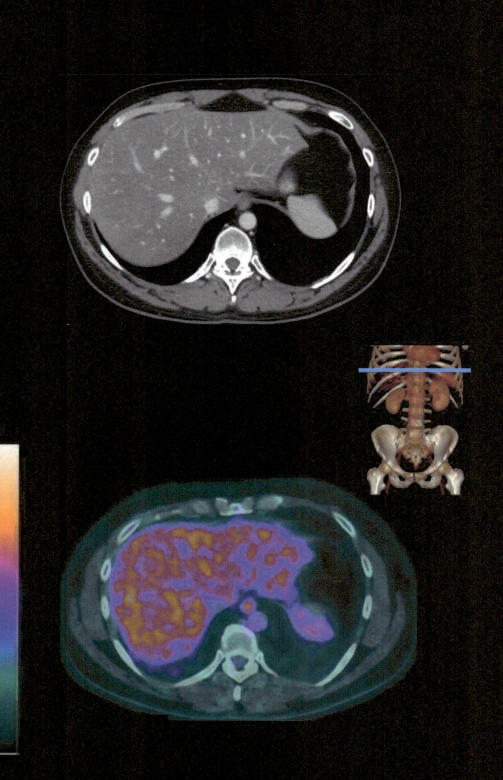

Abdomen and pelvis **Chapter | 2** 175

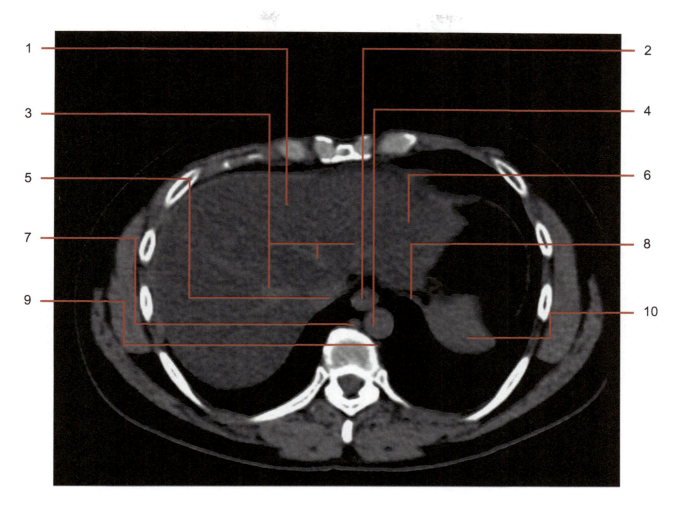

1	Right lobe of liver	6	Left lobe of liver
2	Esophagus	7	Azygos vein
3	Hepatic veins	8	Left hemidiaphragm
4	Thoracic aorta	9	Hemiazygos vein
5	Inferior vena cava (retrohepatic)	10	Spleen

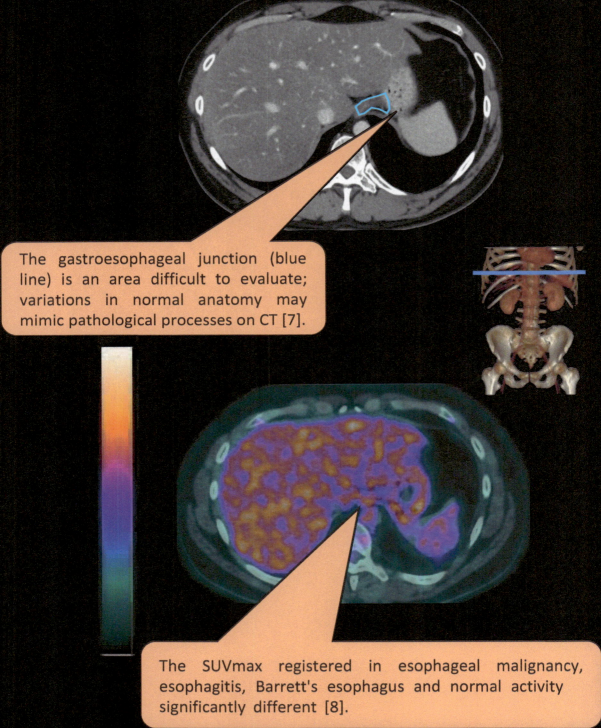

Abdomen and pelvis Chapter | 2 177

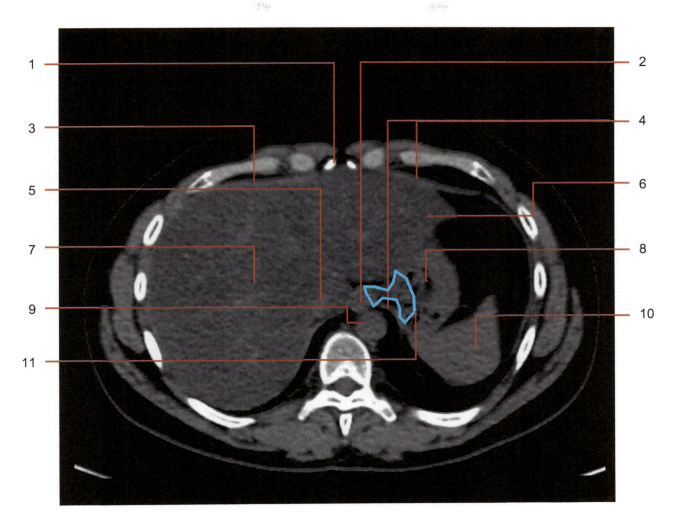

1 Xiphoid process
2 Esophagus
3 Right hemidiaphragm
4 Left hemidiaphragm
5 Inferior vena cava (retrohepatic)
6 Left lobe of liver
7 Right lobe of liver
8 Stomach
9 Thoracic aorta
10 Spleen
11 Gastroesophageal junction

See Refs [7,8]

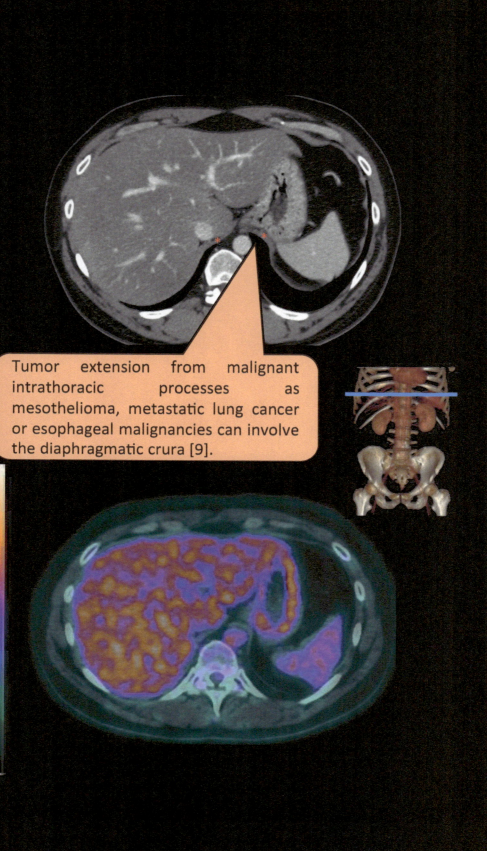

Abdomen and pelvis Chapter | 2 179

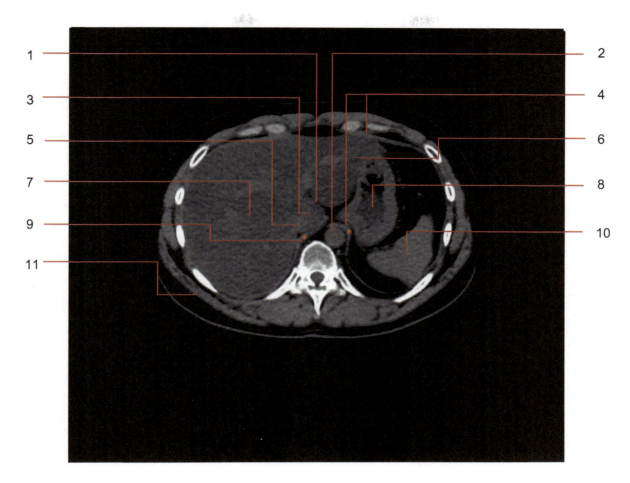

1 Fissure for ligamentum venosum
2 Thoracic aorta
3 Caudate lobe
4 Diaphragm (*crura)
5 Inferior vena cava (retrohepatic)
6 Left lobe of liver
7 Right lobe of liver
8 Stomach
9 Diaphragm (right crus)
10 Spleen
11 Right latissimus dorsi muscle

See Ref [9]

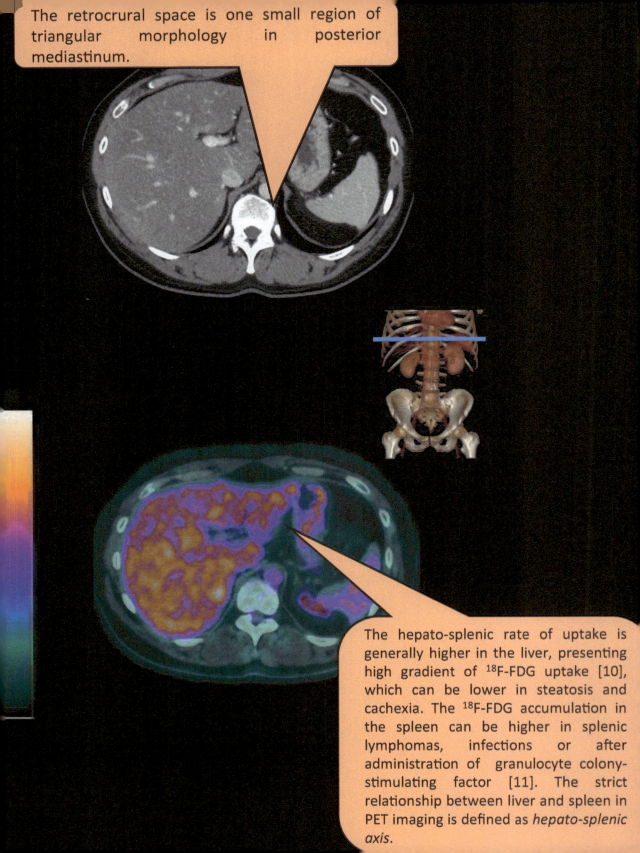

Abdomen and pelvis **Chapter | 2** 181

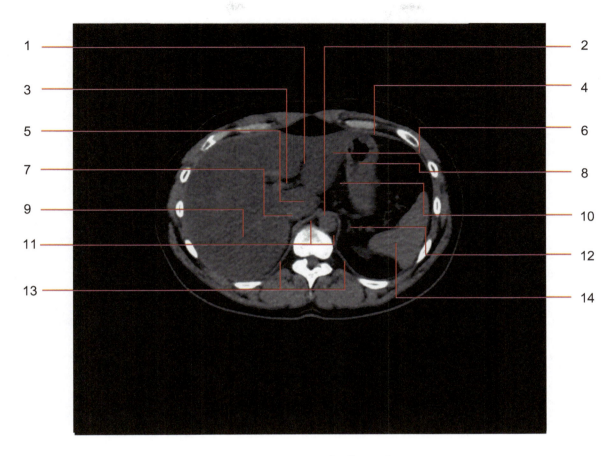

1 Fissure for round ligament
2 Thoracic aorta
3 Transverse fissure
4 Left hemidiaphragm
5 Caudate lobe
6 Left lobe of liver
7 Inferior vena cava (retrohepatic)
8 Stomach
9 Right lobe of liver
10 Gastrohepatic (hepatogastric) ligament
11 Retrocrural space (right and left)
12 Left adrenal gland
13 Diaphragm
14 Spleen

See Refs [10,11]

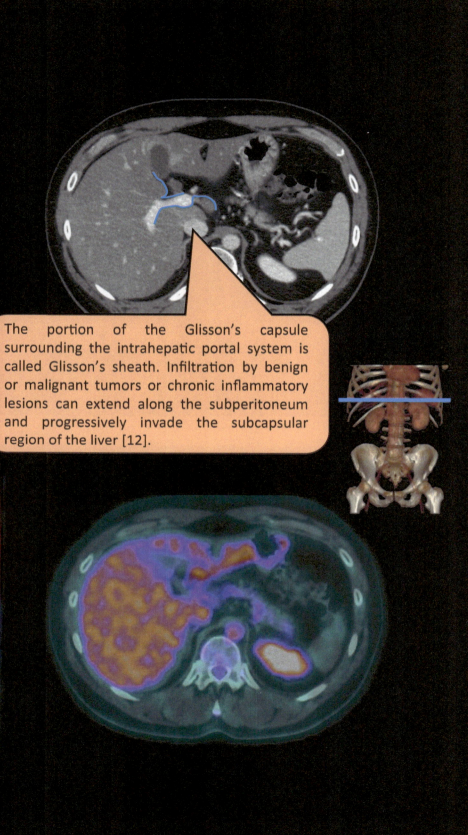

The portion of the Glisson's capsule surrounding the intrahepatic portal system is called Glisson's sheath. Infiltration by benign or malignant tumors or chronic inflammatory lesions can extend along the subperitoneum and progressively invade the subcapsular region of the liver [12].

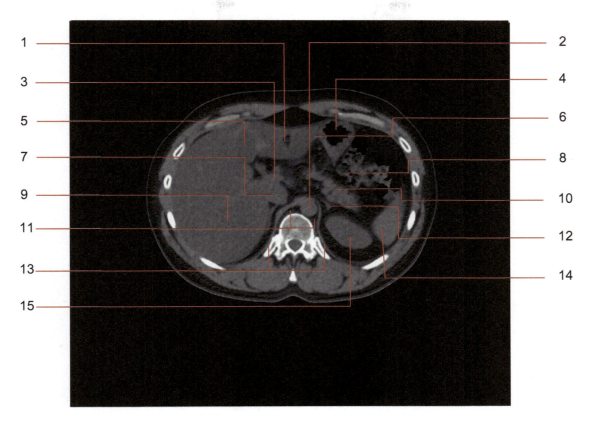

1 Fissure and round ligament
2 Thoracic aorta
3 Hepatic portal vein
4 Stomach
5 Gallbladder
6 Left lobe of liver
7 Inferior vena cava (retrohepatic)
8 Colon (left flexure)
9 Right lobe of liver
10 Pancreas (tail)
11 Retrocrural space (right and left)
12 Left adrenal gland
13 Right and left hemidiaphragm
14 Spleen
15 Left kidney

See Ref [12]

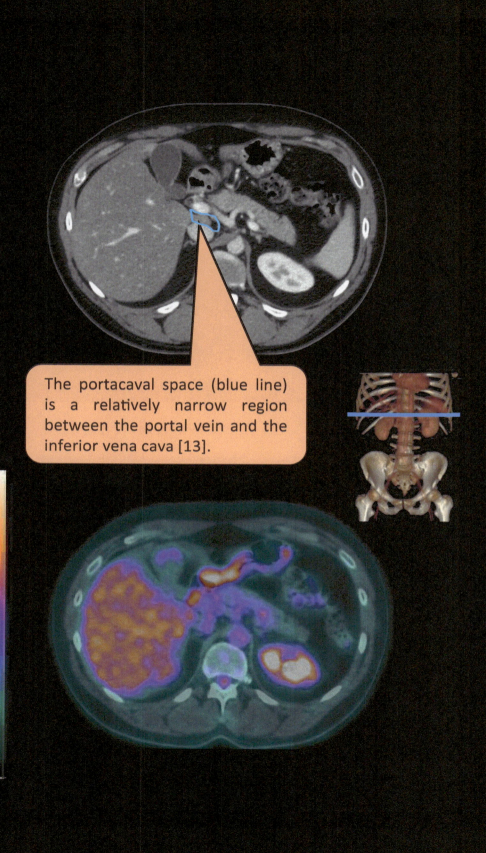

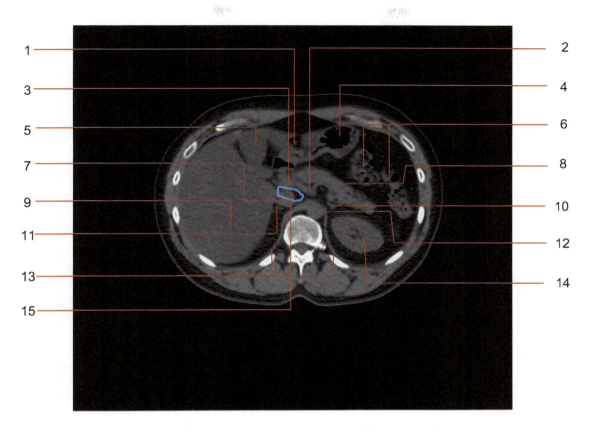

1 Round ligament of liver
2 Celiac trunk (division)
3 Hepatic portal vein
4 Stomach
5 Gallbladder
6 Colon (left flexure)
7 Inferior vena cava (infrahepatic)
8 Pancreas (body–tail)
9 Right lobe of liver
10 Left adrenal gland
11 Right adrenal gland
12 Abdominal aorta
13 Right and left hemidiaphragm
14 Left kidney
15 Portacaval space

See Ref [13]

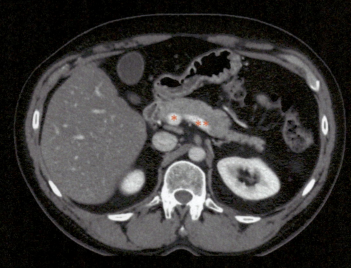
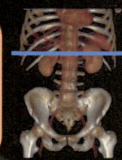
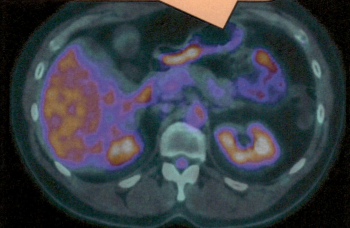

Per os idratation immediately prior of PET/CT may enlarge gastric wall, allowing better delineation of the stomach in PET/CT and CT images, during examination of gastric cancers and lymphomas [14].

Abdomen and pelvis Chapter | 2 187

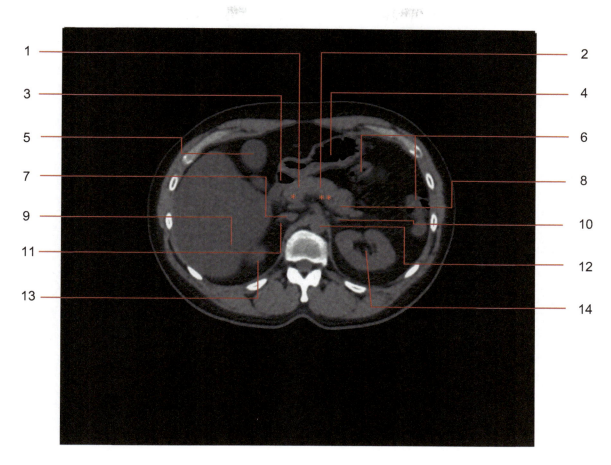

1 Pancreas (head)* hepatic portal vein
2 Pancreas (body) **splenic vein
3 Duodenal bulb
4 Stomach
5 Gallbladder
6 Colon (tranverse and descending)
7 Inferior vena cava (infrahepatic)
8 Pancreas (tail)
9 Right lobe of liver
10 Left adrenal gland
11 Right adrenal gland
12 Thoracic aorta
13 Right kidney
14 Left kidney

See Ref [14]

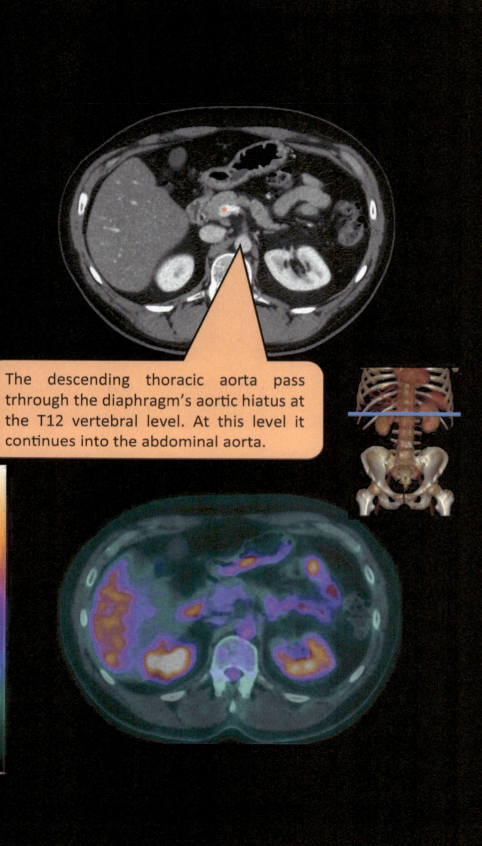

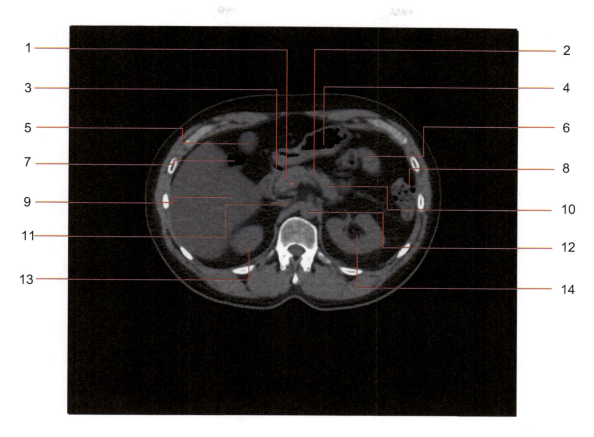

1 Pancreas (head) * superior mesenteric vein
2 Pancreas (body)
3 Duodenum (bulb)
4 Stomach
5 Gallbladder
6 Jejunum
7 Colon (right flexure)
8 Colon (descending)
9 Right lobe of liver
10 Pancreas (tail)
11 Inferior vena cava (infrahepatic)
12 Abdominal aorta
13 Right kidney
14 Left kidney

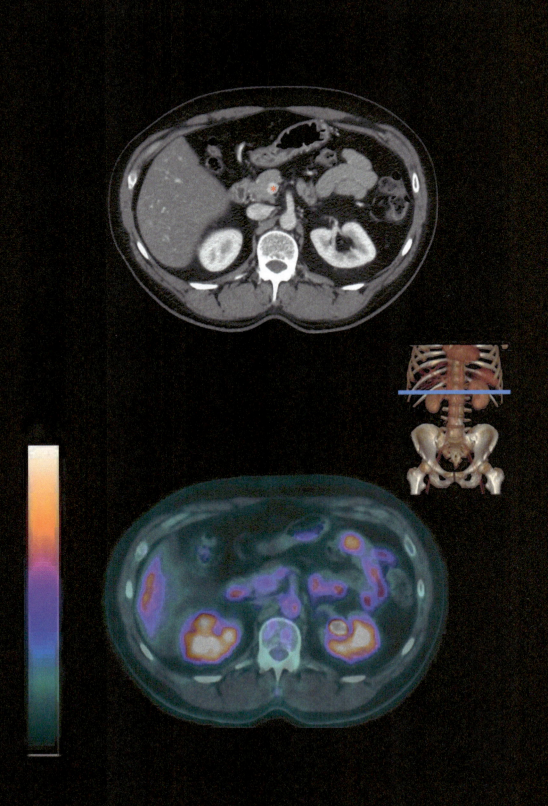

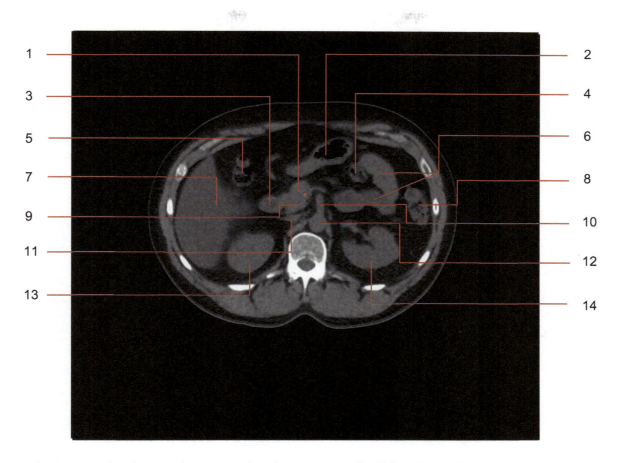

1 Pancreas (head) * superior mesenteric vein
2 Stomach
3 Duodenum (descending part)
4 Colon (transverse)
5 Colon (right flexure)
6 Jejunum
7 Right lobe of liver
8 Colon (descending)
9 Uncinate process of pancreas
10 Superior mesenteric artery
11 Inferior vena cava (infrahepatic)
12 Abdominal aorta
13 Right kidney
14 Left kidney

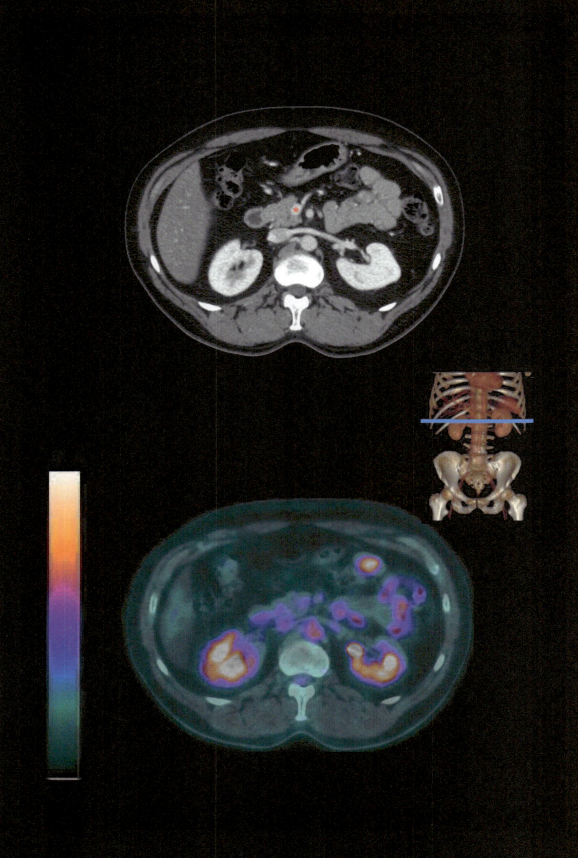

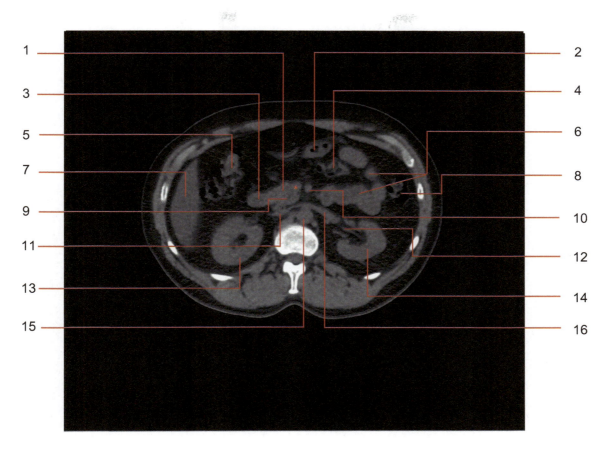

1 Pancreas (head) * superior mesenteric vein
2 Stomach (greater curvature)
3 Duodenum (descending part)
4 Colon (transverse)
5 Colon (right flexure)
6 Jejunum
7 Right lobe of liver
8 Colon (descending)
9 Uncinate process
10 Superior mesenteric artery
11 Inferior vena cava (infrahepatic)
12 Renal pelvis
13 Right kidney
14 Left kidney
15 Abdominal aorta
16 Left renal vein

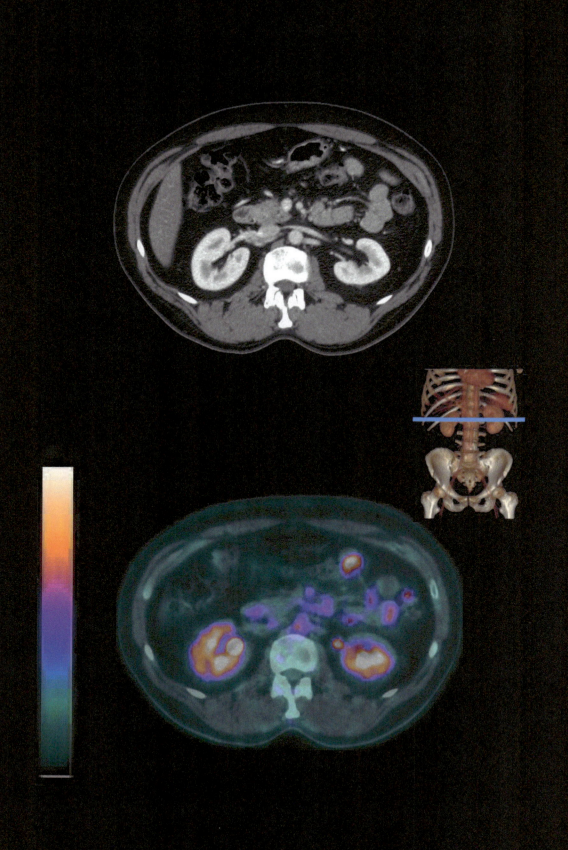

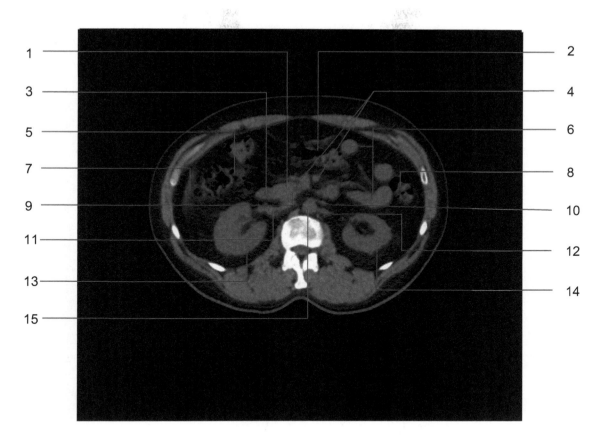

1 Pancreas (head)
2 Colon (transverse)
3 Duodenum (descending part)
4 Superior mesenteric vein (right) and artery (left)
5 Colon (right flexure)
6 Jejunum
7 Liver (inferior margin of right lobe)
8 Colon (descending)
9 Uncinate process of pancreas
10 Left renal vein
11 Inferior vena cava (infrahepatic)
12 Left renal artery
13 Right kidney
14 Left kidney
15 Abdominal aorta

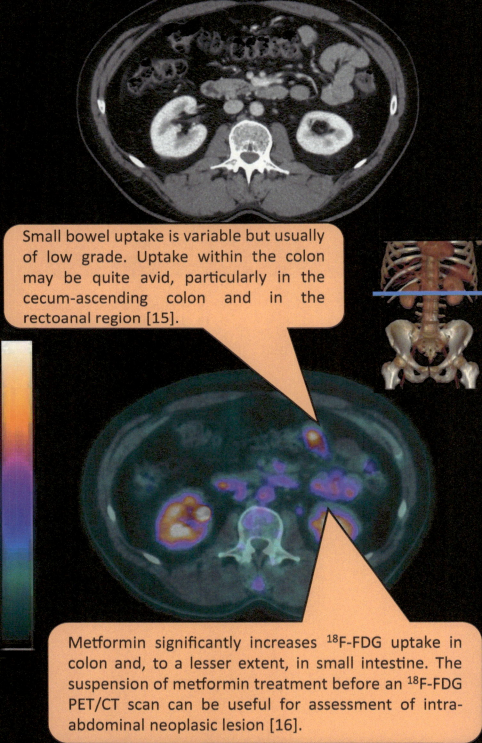

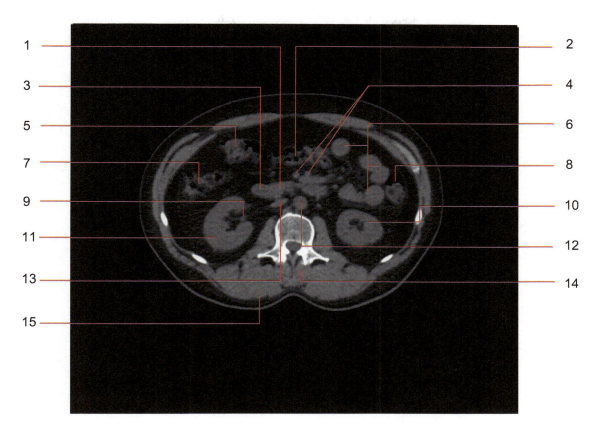

1 Uncinate process of pancreas
2 Colon (transverse)
3 Duodenum (descending part)
4 Superior mesenteric vein (right) and artery (left)
5 Colon (transverse)
6 Jejunum
7 Colon (ascending)
8 Colon (descending)
9 Renal pelvis
10 Left kidney
11 Right kidney
12 Abdominal aorta
13 Inferior vena cava (infrahepatic)
14 Left spinalis muscle
15 Right longissimus dorsi muscle

See Refs [15,16]

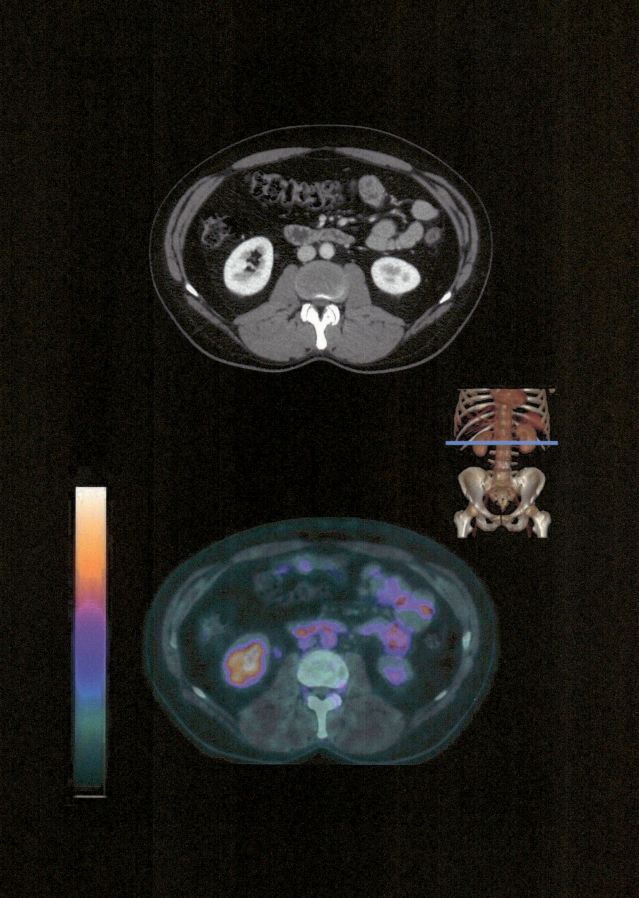

Abdomen and pelvis Chapter | 2   199

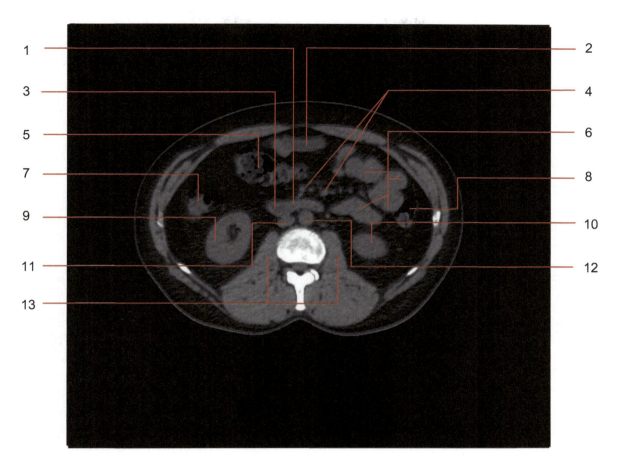

1 Duodenum (horizontal part)
2 Jejunum
3 Duodenum (inferior flexure)
4 Superior mesenteric vein (right) and artery (left)
5 Colon (transverse)
6 Jejunum
7 Colon (ascending)
8 Colon (descending)
9 Right kidney
10 Left kidney
11 Inferior vena cava (infrahepatic)
12 Abdominal aorta
13 Psoas major muscles

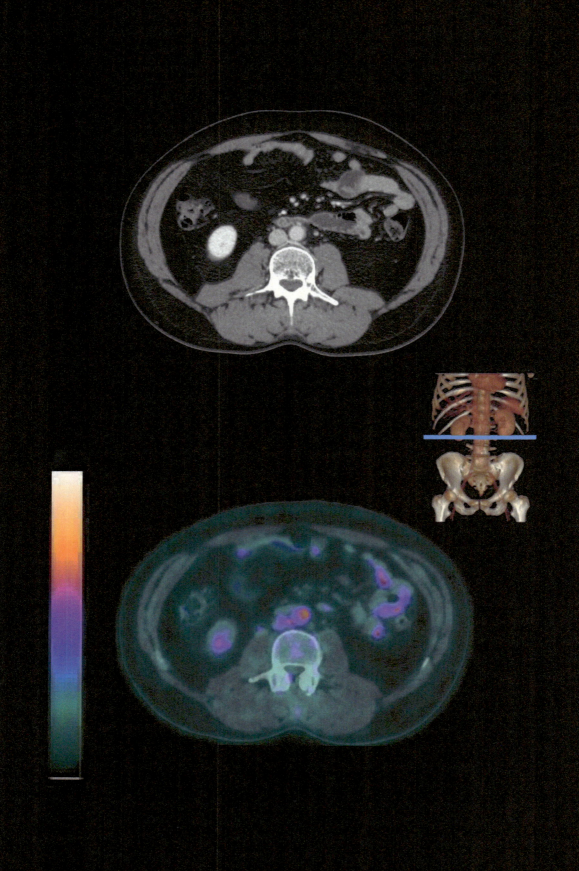

Abdomen and pelvis Chapter | 2   201

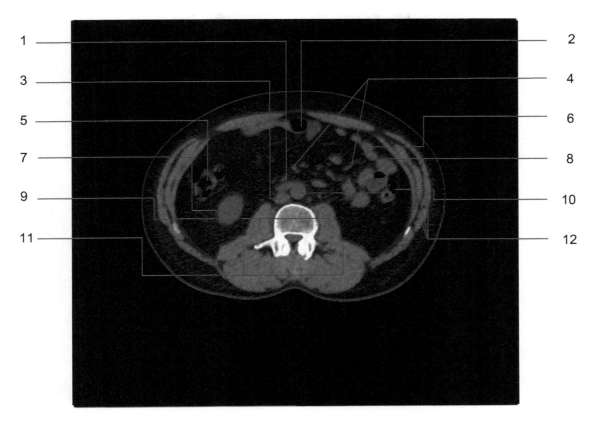

1 Duodenum (horizontal part)
2 Jejunum
3 Inferior vena cava (infrahepatic)
4 Ileal and jejunal arteries and veins
5 Colon (ascending)
6 Jejunum
7 Right kidney
8 Abdominal aorta
9 Psoas major muscles
10 Colon (descending)
11 Quadratus lumborum muscles
12 Abdominal muscles (antero—lateral wall)

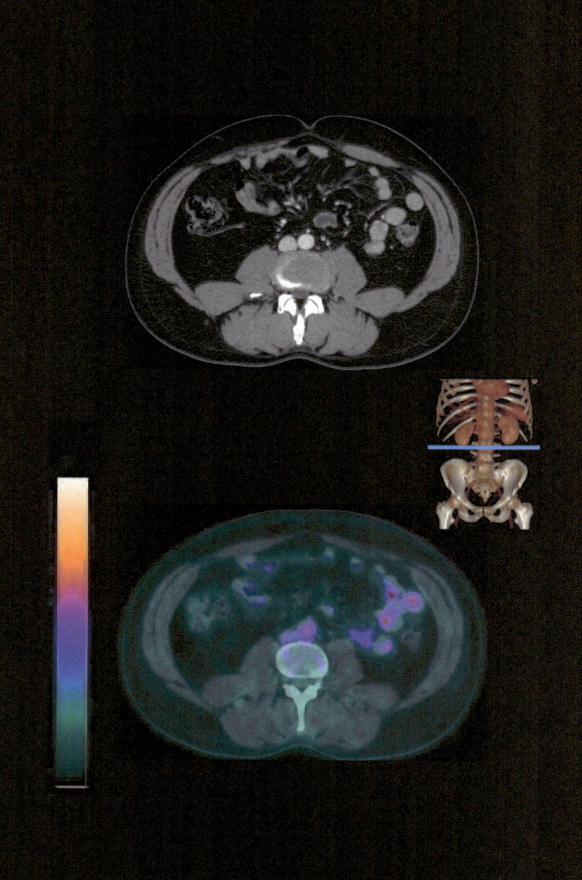

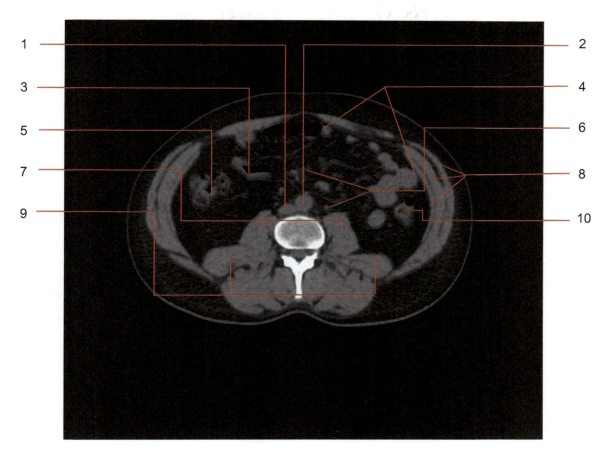

1 Inferior vena cava (infrahepatic)
2 Abdominal aorta
3 Ileum
4 Jejunum
5 Colon (ascending)
6 Ileal and jejunal arteries and veins
7 Psoas major muscles
8 Abdominal muscles (antero–lateral wall)
9 Quadratus lumborum muscles
10 Colon (descending)

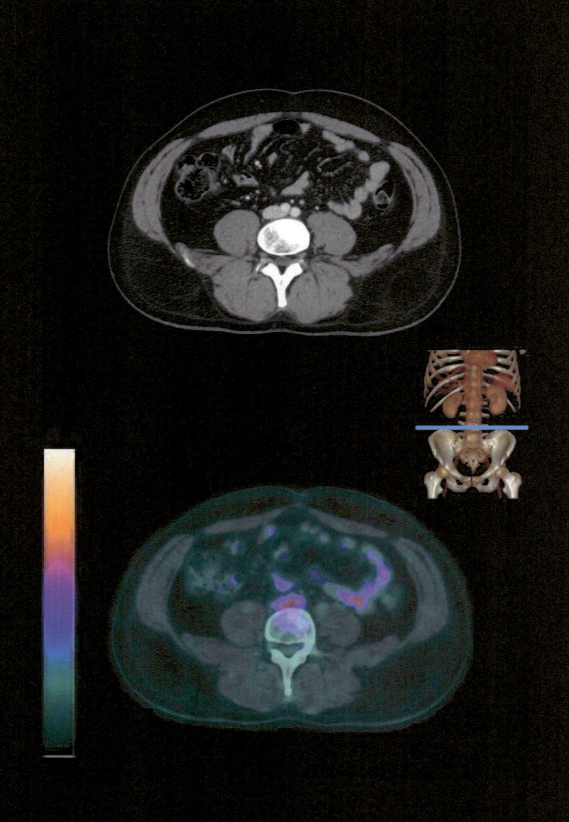

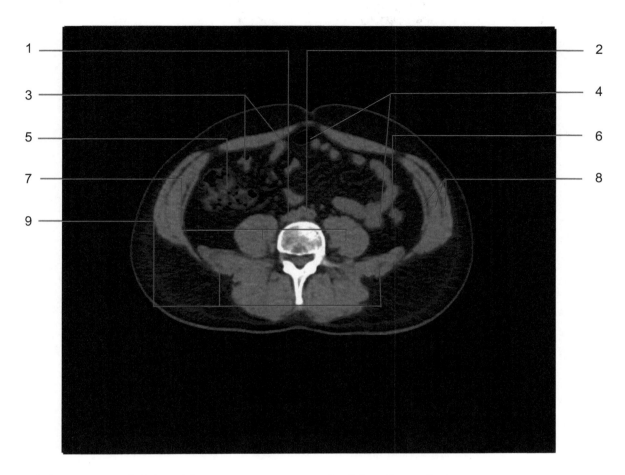

1 Inferior vena cava (infrahepatic)
2 Aorta bifurcation («*carrefour*»)
3 Ileum
4 Jejunum
5 Colon (ascending)
6 Colon (descending)
7 Psoas major muscles
8 Abdominal muscles (antero−lateral wall)
9 Quadratus lumborum muscles

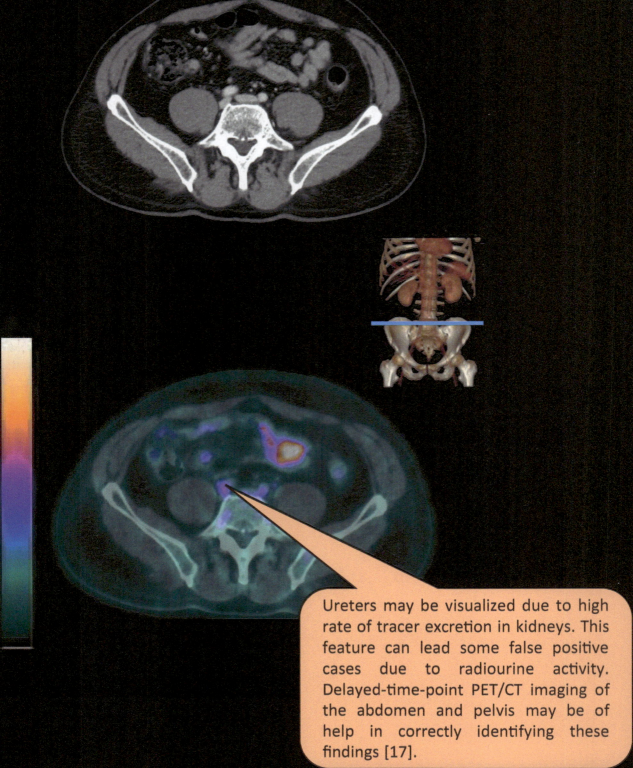

Abdomen and pelvis Chapter | 2 | 207

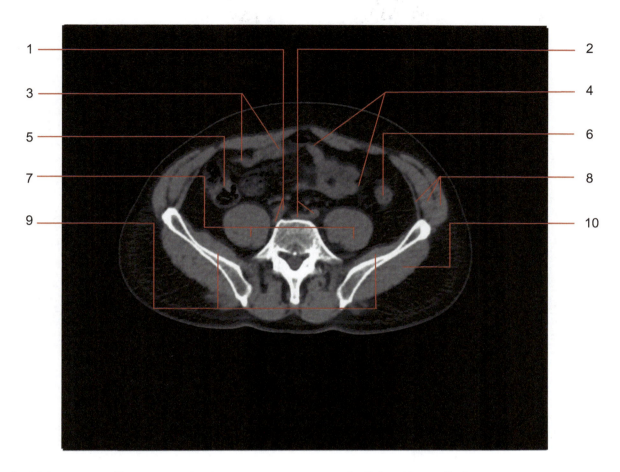

1 Right common iliac artery and vein
2 Left common iliac artery and vein
3 Ileum
4 Jejunum
5 Cecum
6 Colon (descending)
7 Psoas major muscles
8 Abdominal muscles (antero−lateral wall)
9 Iliacus muscles
10 Left gluteus muscle

See Ref [17]

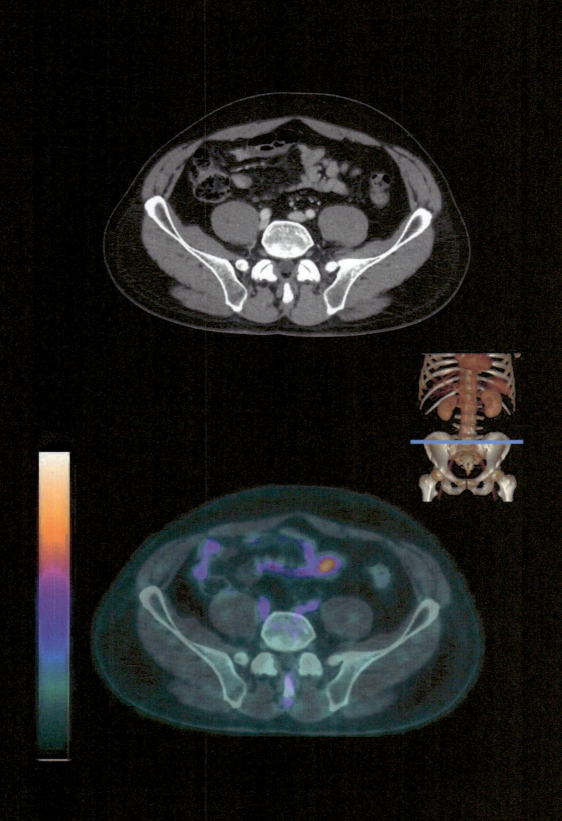

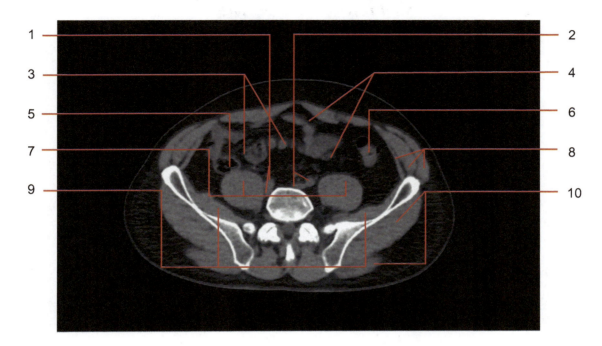

1 Right common iliac artery and vein
2 Left common iliac artery and vein
3 Ileum
4 Jejunum
5 Cecum
6 Colon (descending)
7 Psoas major muscles
8 Abdominal muscles (antero−lateral wall)
9 Iliacus muscles
10 Left gluteus muscles

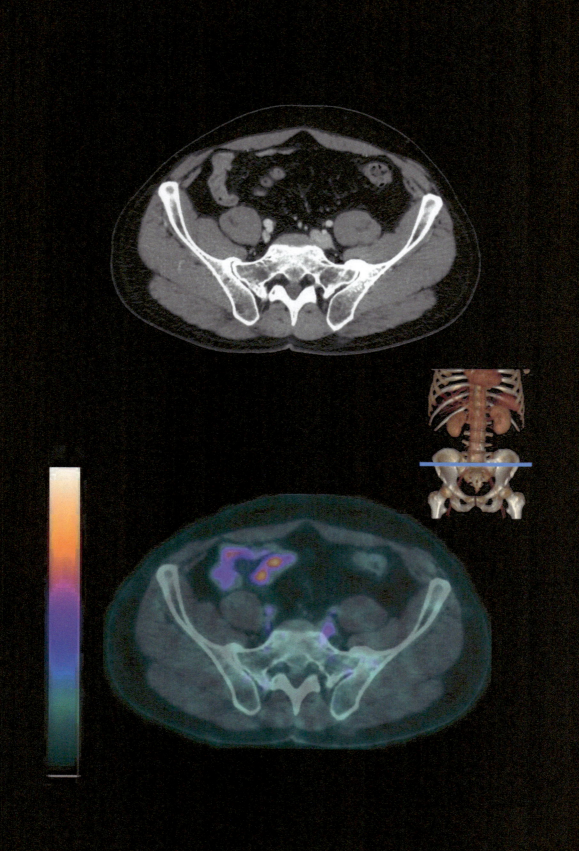

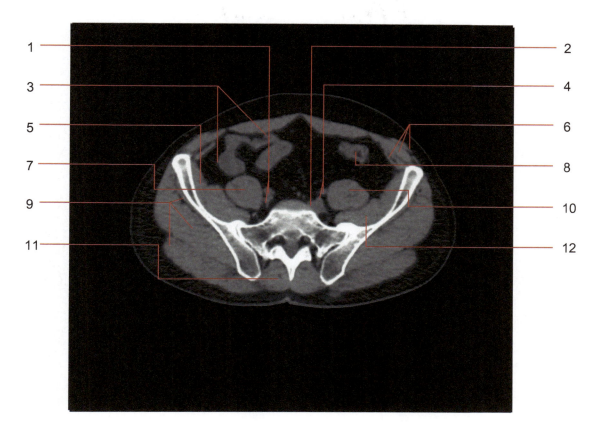

1 Right common iliac artery and vein
2 Left common iliac vein
3 Ileum
4 Left internal and external iliac artery
5 Right iliacus muscle
6 Abdominal muscles (antero−lateral wall)
7 Right psoas major muscle
8 Colon (descending)
9 Right gluteus muscles (minimus, medius, maximus)
10 Left psoas major muscle
11 Right spinalis muscle
12 Left iliacus muscle

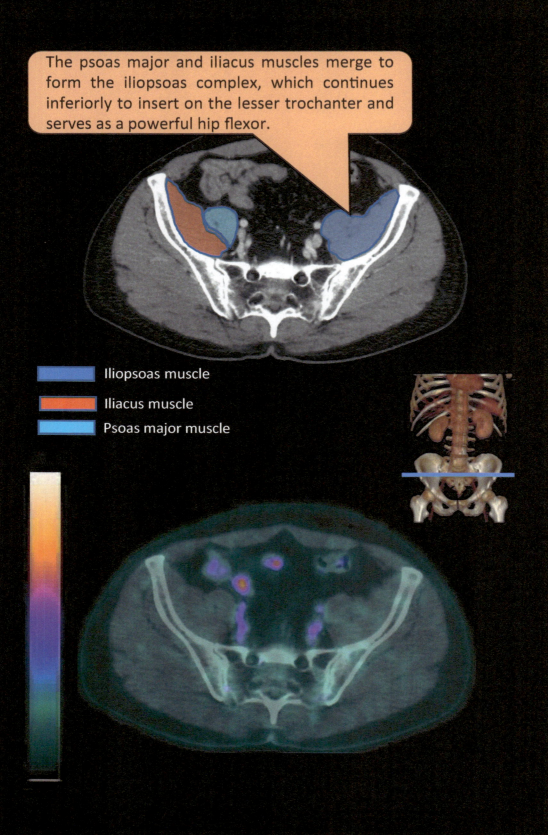

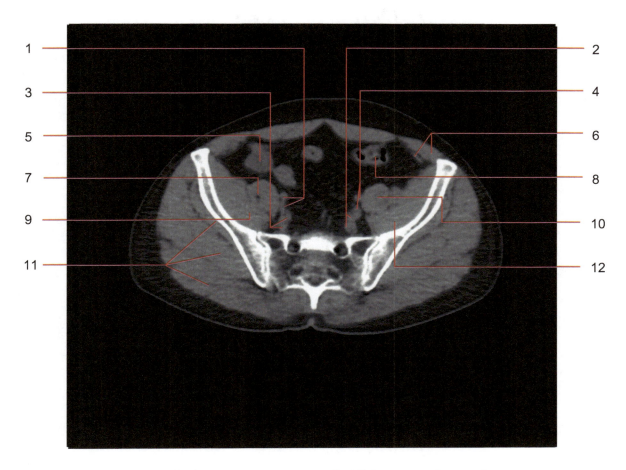

1 Right external iliac artery and vein
2 Left internal iliac artery and vein
3 Right internal iliac artery and vein
4 Left external iliac artery and vein
5 Ileum
6 Abdominal muscles (antero—lateral wall)
7 Right psoas major muscle
8 Colon (sigmoid)
9 Right iliacus muscle
10 Left psoas major muscle
11 Right gluteus muscles (minimus, medius, maximus)
12 Left iliacus muscle

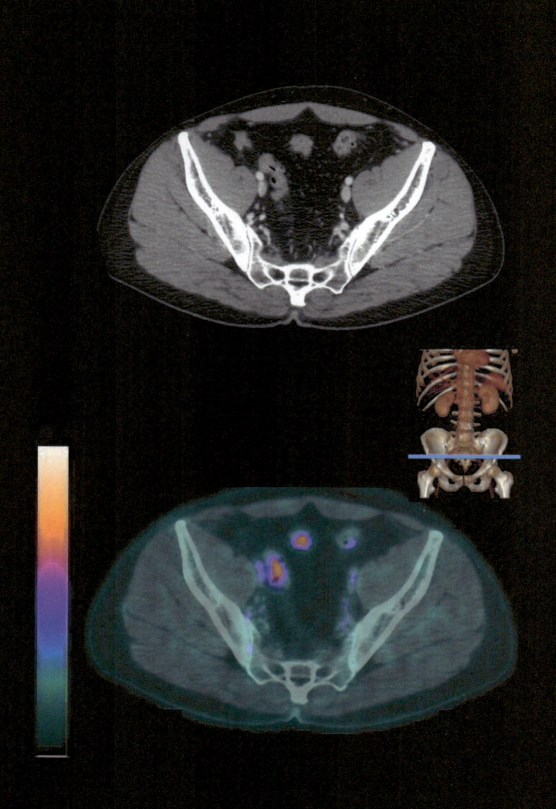

Abdomen and pelvis Chapter | 2   215

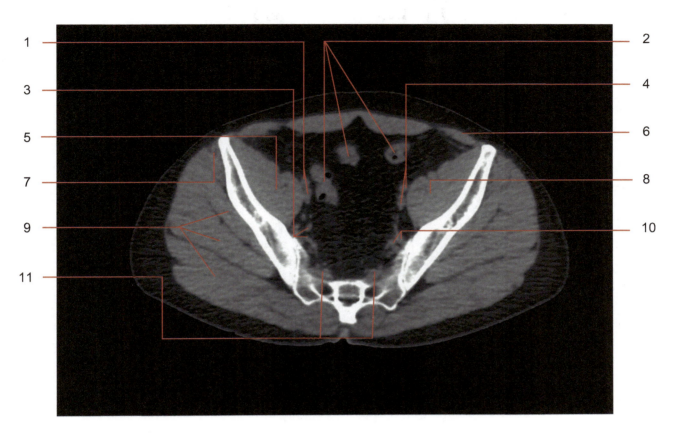

1 Right external iliac artery and vein
2 Colon (sigmoid)
3 Right internal iliac artery and vein
4 Left external iliac artery and vein
5 Right iliopsoas muscle
6 Abdominal muscles (antero−lateral wall)
7 Right sartorius muscle
8 Left iliopsoas muscle
9 Right gluteus muscles (minimus, medius, maximus)
10 Left internal iliac artery and vein
11 Piriformis muscles

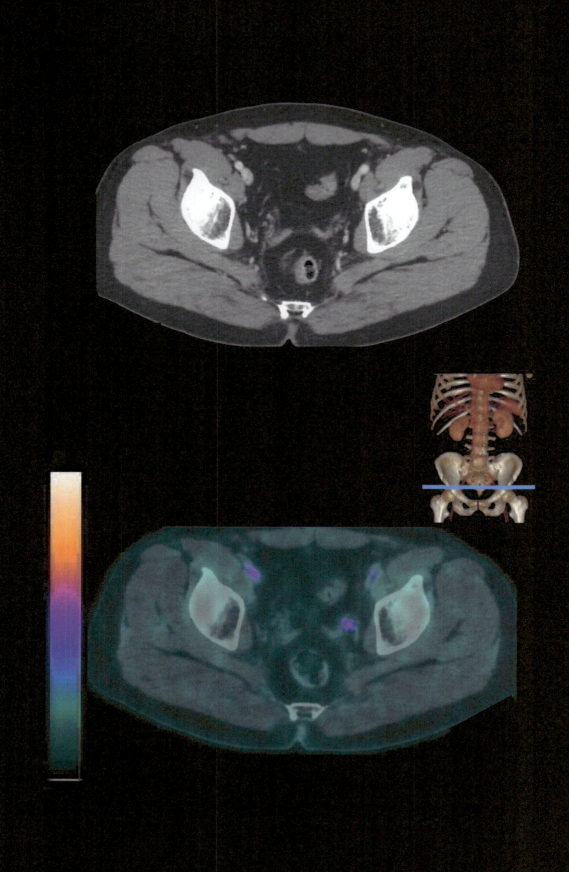

Abdomen and pelvis Chapter | 2   217

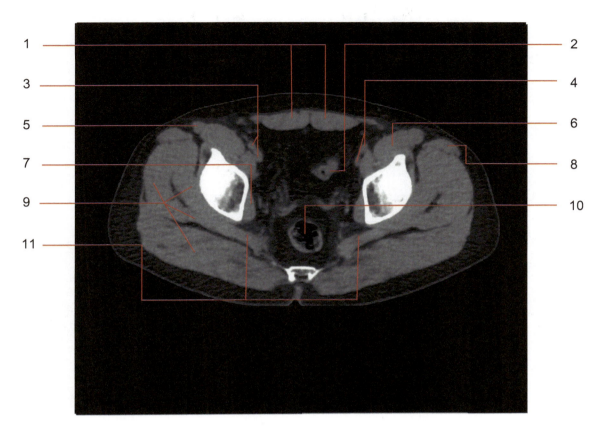

1 Rectus abdominis muscles
2 Colon (sigmoid)
3 Right external iliac artery and vein
4 Left external iliac artery and vein
5 Right sartorius muscle
6 Left iliopsoas muscle
7 Right obturator internus muscle
8 Left tensor fasciae latae muscle
9 Right gluteus muscles (minimus, medius, maximus)
10 Rectum (ampulla)
11 Piriformis muscles

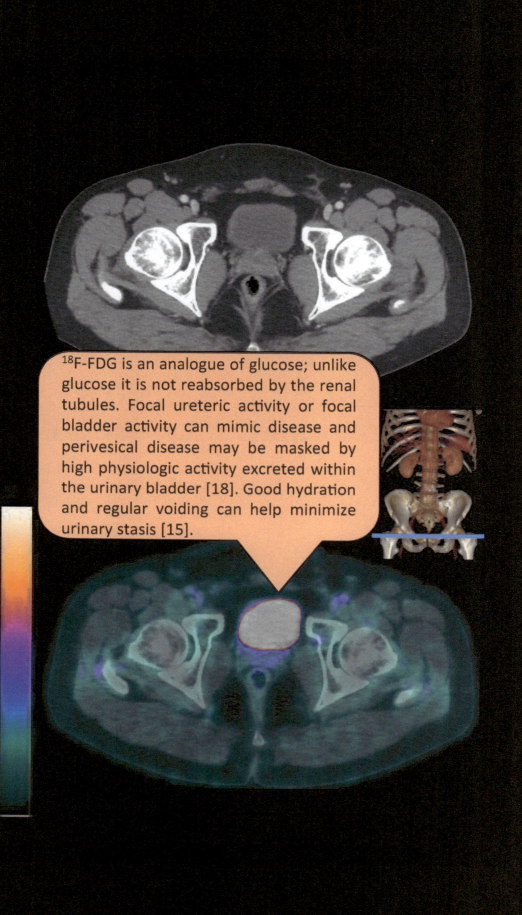

^{18}F-FDG is an analogue of glucose; unlike glucose it is not reabsorbed by the renal tubules. Focal ureteric activity or focal bladder activity can mimic disease and perivesical disease may be masked by high physiologic activity excreted within the urinary bladder [18]. Good hydration and regular voiding can help minimize urinary stasis [15].

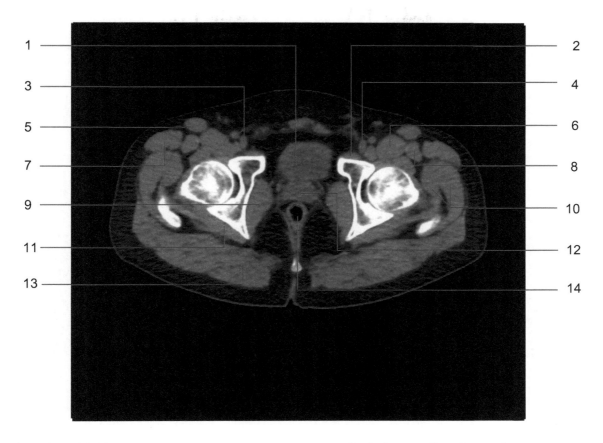

1 Urinary bladder
2 Left obturator externus muscle
3 Right external iliac artery and vein
4 Left external iliac artery and vein
5 Right sartorius muscle
6 Left iliopsoas muscle
7 Right tensor fasciae latae muscle
8 Left rectus femoris muscle
9 Right obturator internus muscle
10 Left gluteus muscles
11 Right piriformis muscle
12 Prostate (base)
13 Right gluteus muscle
14 Rectum

See Refs [15,18]

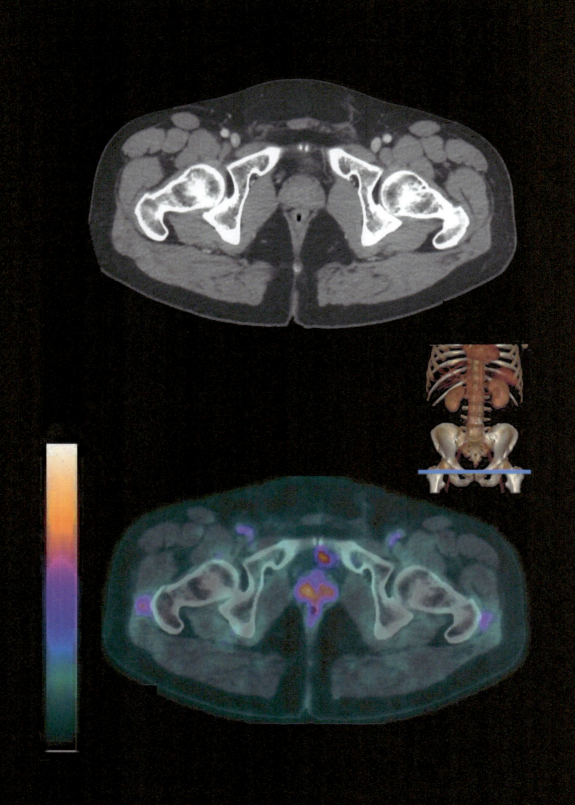

Abdomen and pelvis Chapter | 2   221

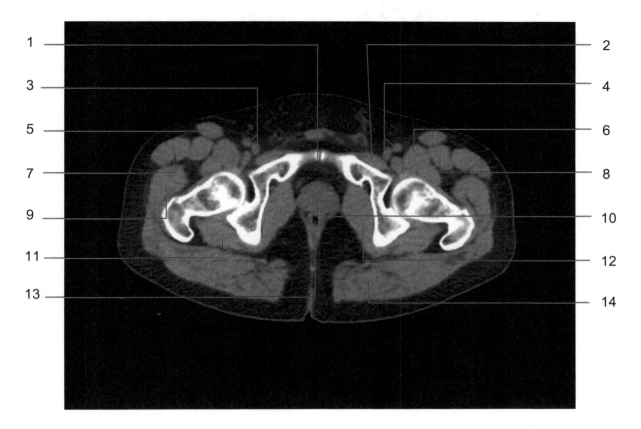

1 Pubic simphysis
2 Left obturator externus muscle
3 Right femoral artery and vein
4 Left femoral artery and vein
5 Right sartorius muscle
6 Left iliopsoas muscle
7 Right tensor fasciae latae muscle
8 Left rectus femoris muscle
9 Right gluteus muscle
10 Prostate (midgland)
11 Right piriformis muscle
12 Left obturator internus muscle
13 Anal canal
14 Left gluteus muscles

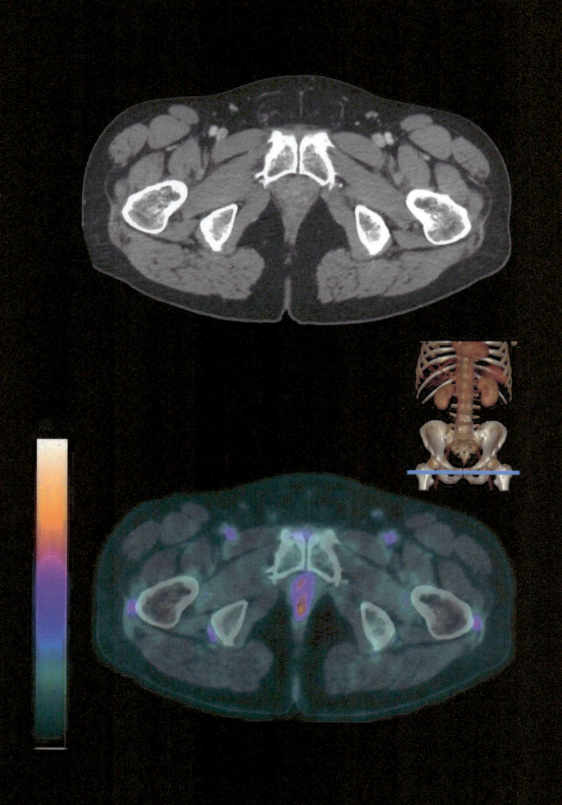

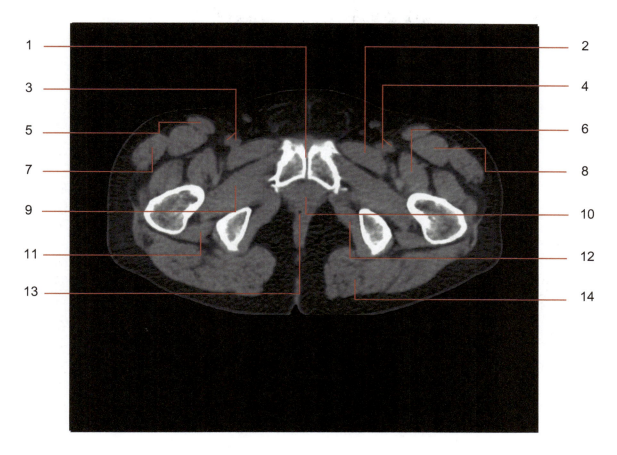

1 Pubic simphysis
2 Left pectineus muscle
3 Right femoral artery and vein
4 Left femoral artery and vein
5 Right sartorius muscle
6 Left iliopsoas muscle
7 Right tensor fasciae latae muscle
8 Left rectus femoris muscle
9 Right obturator externus muscle
10 Prostate (apex)
11 Right quadratus femoris muscle
12 Left obturator internus muscle
13 Anal canal
14 Left gluteus muscle

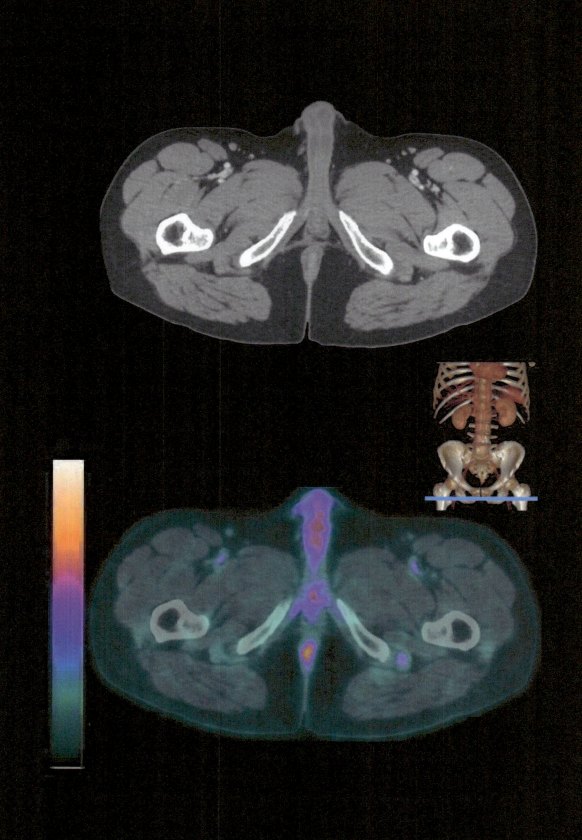

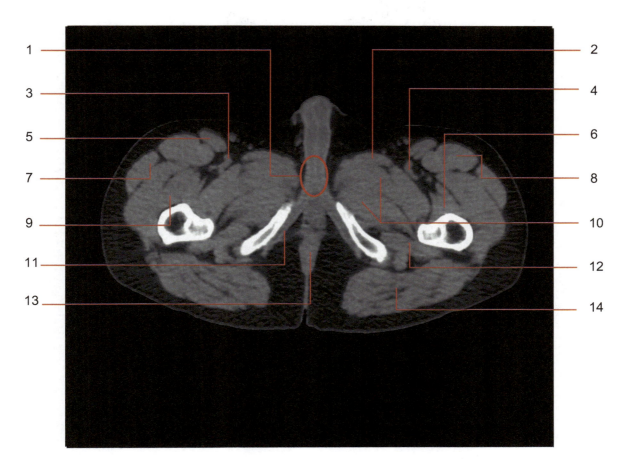

1 Root of penis
2 Adductor longus muscle
3 Right femoral artery and vein
4 Left femoral artery and vein
5 Right sartorius muscle
6 Left iliopsoas muscle
7 Right tensor fasciae latae muscle
8 Left rectus femoris muscle
9 Right gluteus muscle
10 Left adductor brevis and magnus muscles
11 Right ischiocavernosus muscle
12 Left quadratus femoris muscle
13 Anus
14 Left gluteus muscle

2.1.2 Sagittal

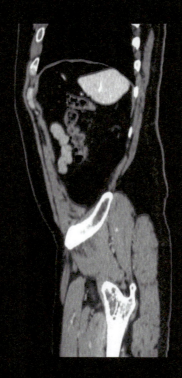
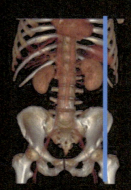
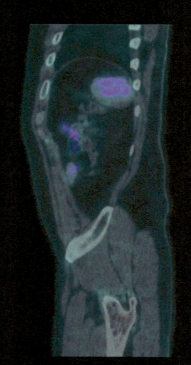

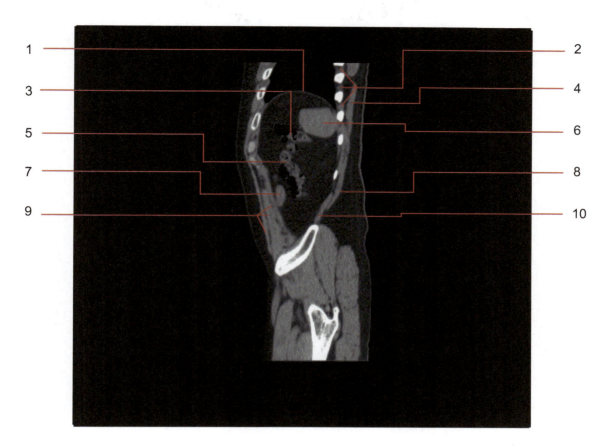

1 Left hemidiaphragm
2 Intercostalis muscles
3 Colon (left flexure)
4 Latissimus dorsi muscle
5 Colon (descending)
6 Spleen
7 Jejunum
8 Latissimus dorsi muscle
9 Abdominal muscles (antero − lateral wall)
10 Quadratus lumborum muscle

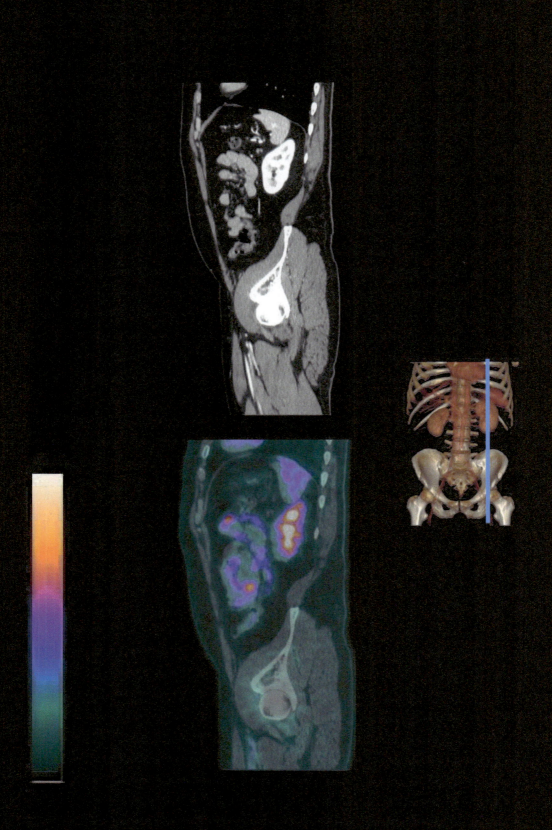

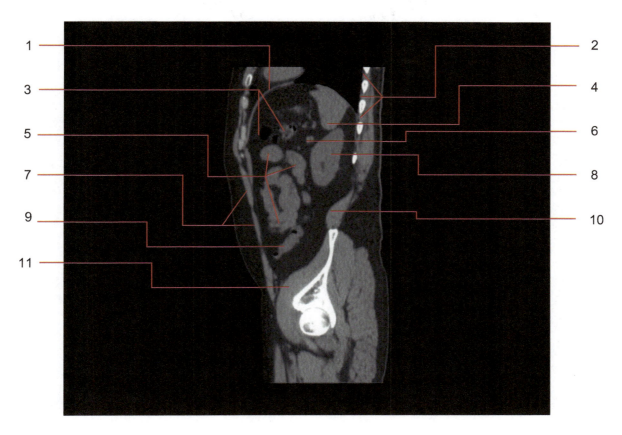

1 Left hemidiaphragm
2 Intercostalis muscles
3 Colon (left flexure, transverse)
4 Spleen
5 Jejunum
6 Pancreas (tail)
7 Rectus abdominis muscle
8 Left kidney
9 Colon (descending)
10 Quadratus lumborum muscle
11 Iliopsoas muscle

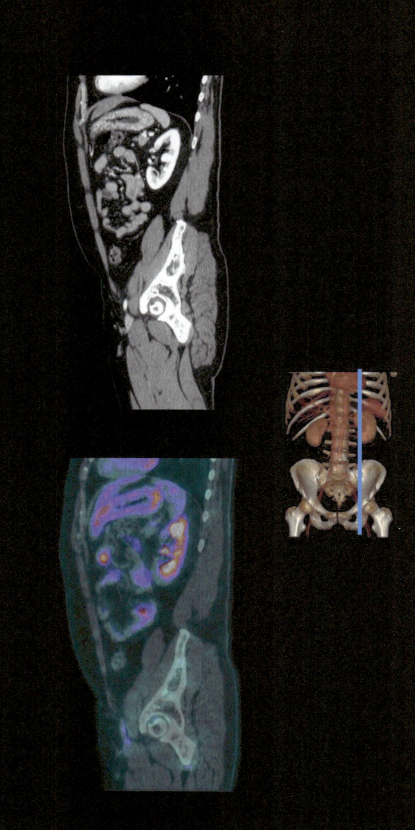

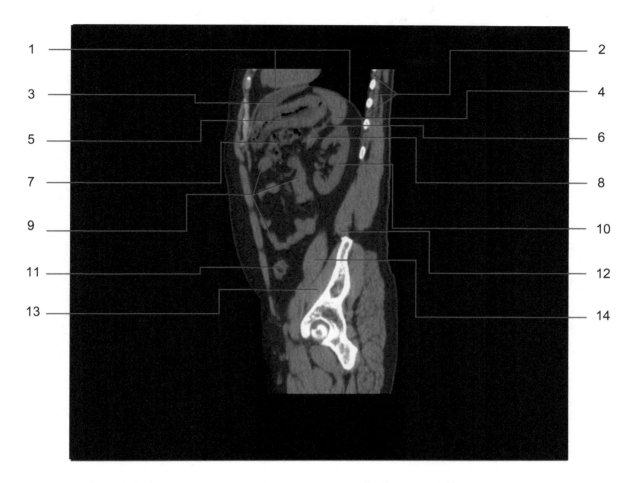

1 Left hemidiaphragm
2 Intercostalis muscles
3 Left lobe of liver
4 Spleen
5 Stomach
6 Left adrenal gland
7 Colon (transverse)
8 Pancreas (tail)
9 Jejunum
10 Left kidney
11 Colon (sigmoid)
12 Quadratus lumborum muscle
13 Iliacus muscle
14 Psoas major muscle

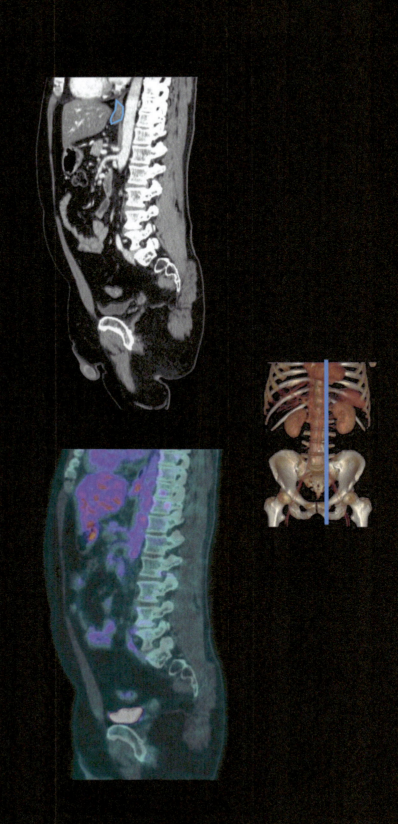

Abdomen and pelvis Chapter | 2

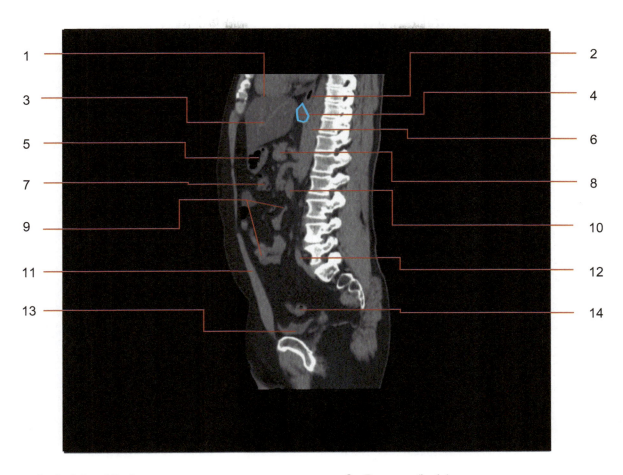

1 Left hemidiaphragm
2 Esophagus
3 Left lobe of liver
4 Gastroesophageal junction
5 Stomach
6 Abdominal aorta
7 Colon (transverse)
8 Pancreas (body)
9 Jejunum
10 Duodenum (horizontal part)
11 Rectus abdominis muscle
12 Left common iliac artery and vein
13 Urinary bladder
14 Colon (sigmoid)

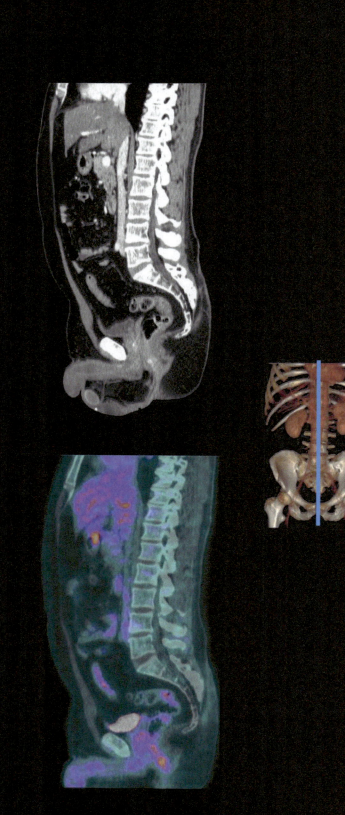

Abdomen and pelvis Chapter | 2   235

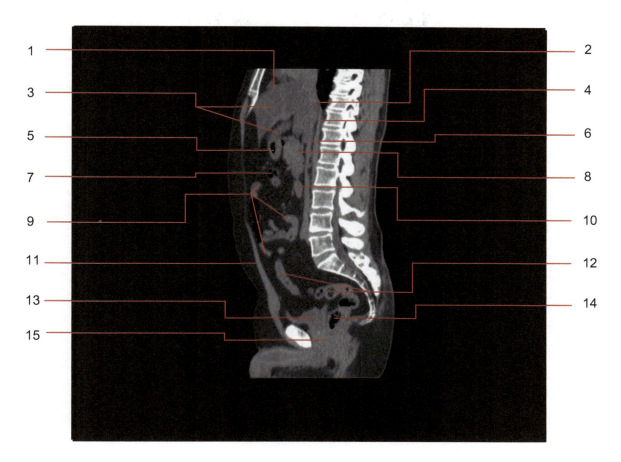

1 Right hemidiaphragm
2 Inferior vena cava (retrohepatic)
3 Left lobe of liver
4 Diaphragm (crus)
5 Stomach
6 Inferior vena cava (infrahepatic)
7 Colon (transverse)
8 Pancreas (head)
9 Jejunum
10 Duodenum (horizontal part)
11 Rectus abdominis muscle
12 Colon (sigmoid)
13 Urinary bladder
14 Rectum
15 Anal canal

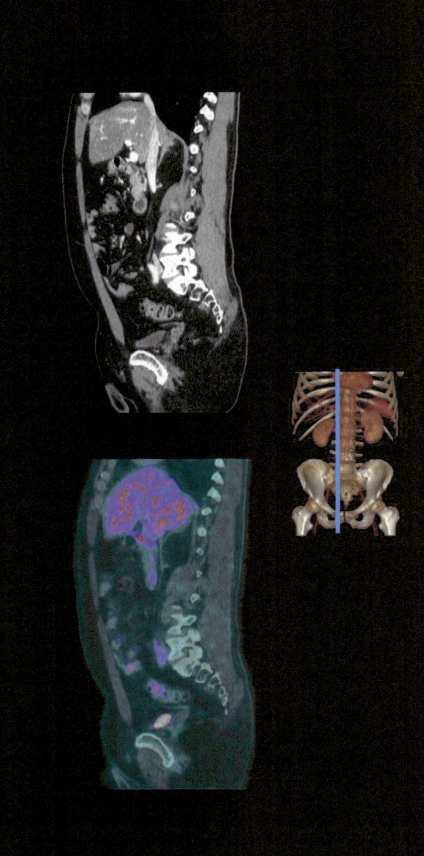

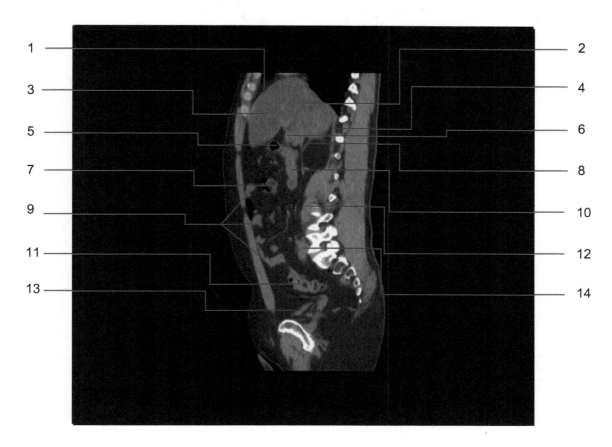

1 Right hemidiaphragm
2 Inferior vena cava (retrohepatic)
3 Right lobe of liver
4 Diaphragm (crura)
5 Duodenum (bulb)
6 Hepatic portal vein
7 Colon (transverse)
8 Right adrenal gland
9 Ileum
10 Duodenum (descending)
11 Colon (sigmoid)
12 Psoas major muscle
13 Urinary bladder
14 Right common iliac artery and vein

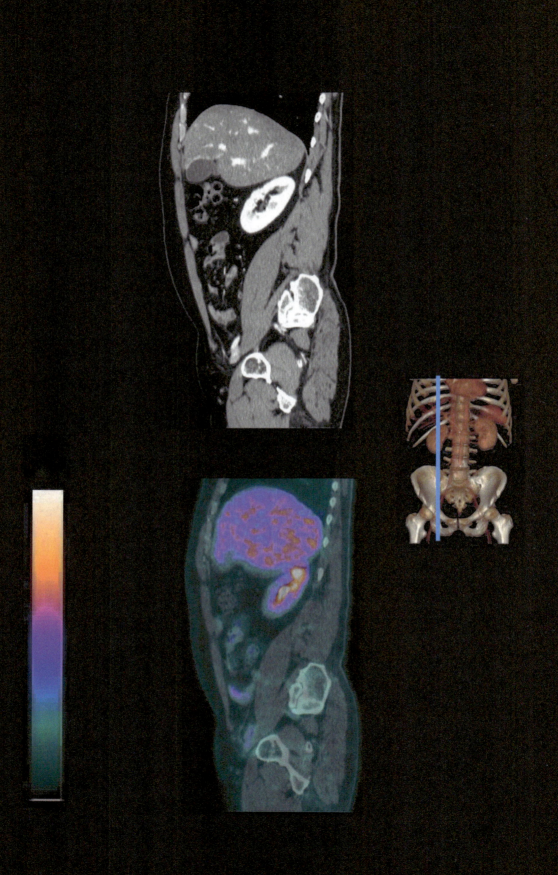

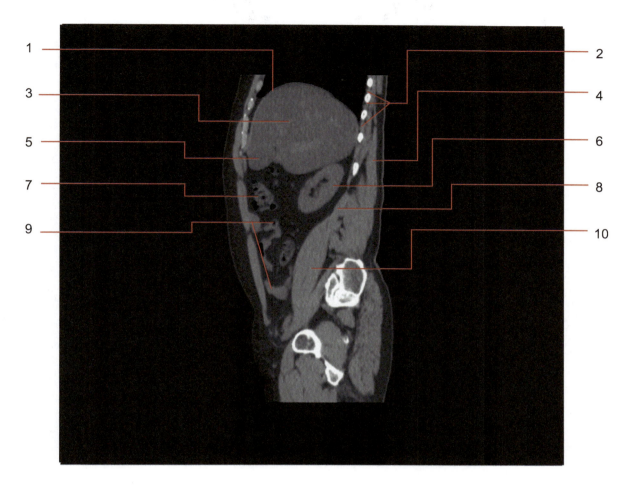

1 Right hemidiaphragm
2 Intercostalis muscles
3 Right lobe of liver
4 Latissimus dorsi muscle
5 Gallbladder
6 Right kidney
7 Colon (transverse)
8 Quadratus lumborum muscle
9 Ileum
10 Psoas major muscle

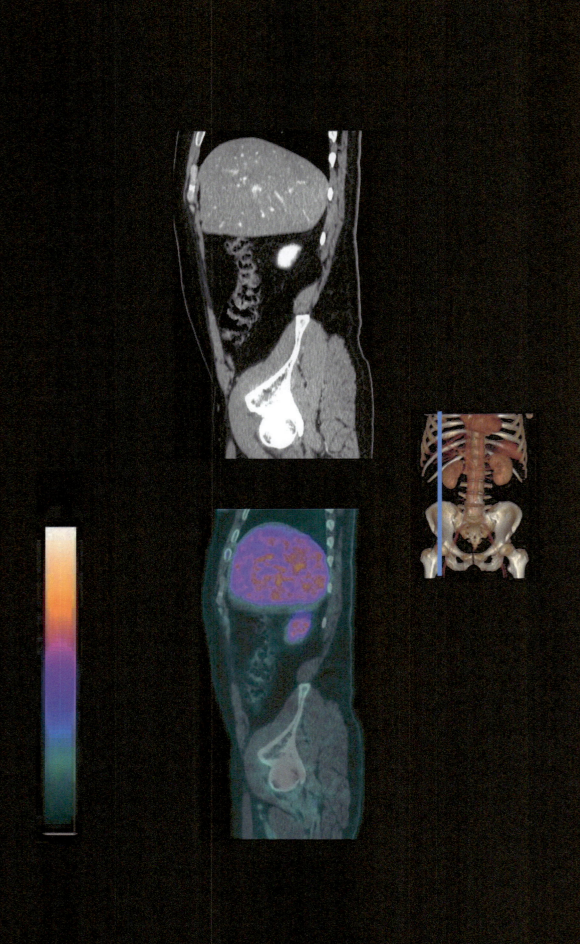

Abdomen and pelvis **Chapter | 2** **241**

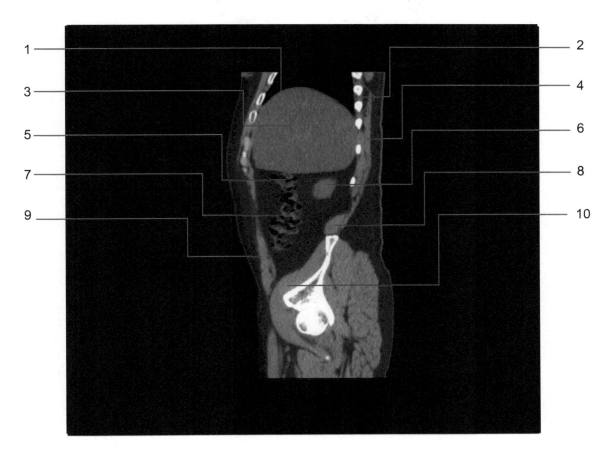

1 Right hemidiaphragm
2 Intercostalis muscles
3 Right lobe of liver
4 Latissimus dorsi muscle
5 Colon (right flexure)
6 Right kidney
7 Colon (ascending)
8 Quadratus lumborum muscle
9 Abdominal muscles (antero–lateral wall)
10 Iliopsoas muscle

2.1.3 Coronal

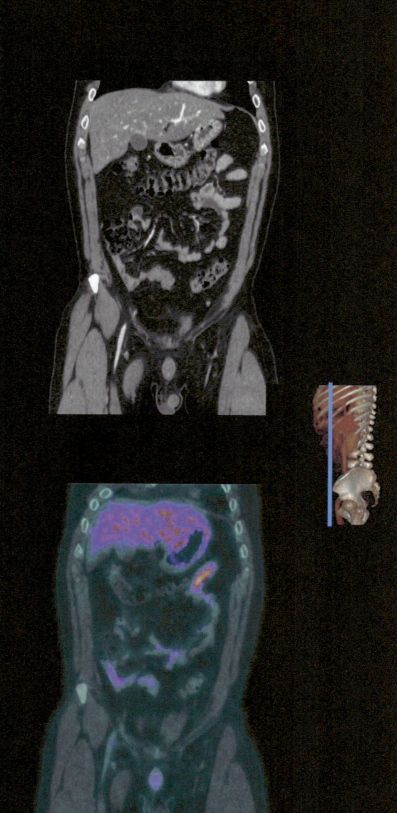

Abdomen and pelvis Chapter | 2   243

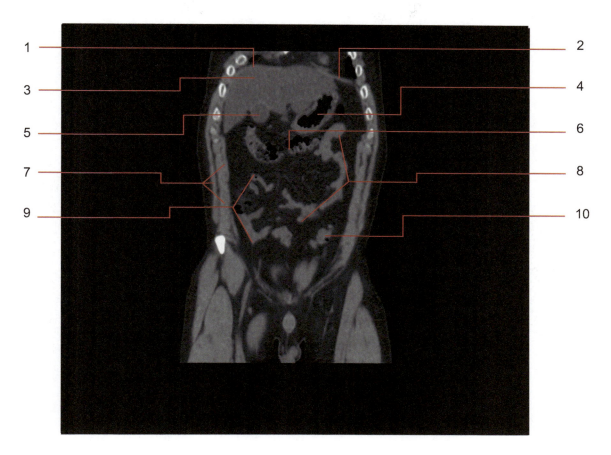

1 Right hemidiaphragm
2 Left hemidiaphragm
3 Liver
4 Stomach
5 Gallbladder
6 Colon (transverse)
7 Abdominal muscles (antero−lateral wall)
8 Jejunum
9 Ileum
10 Colon (descending)

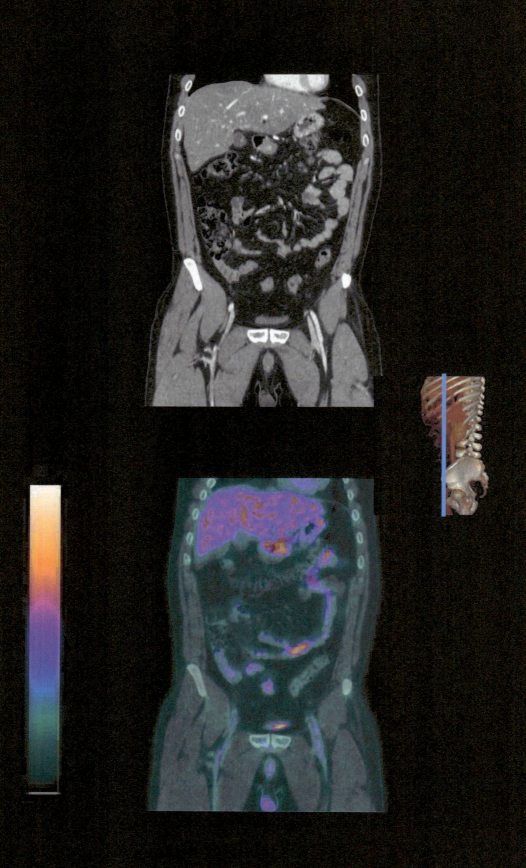

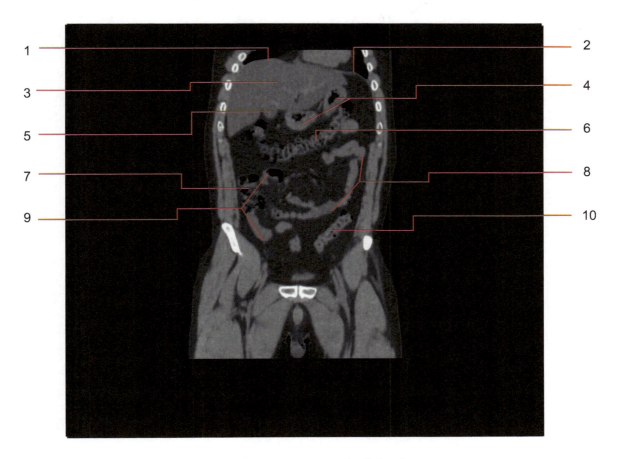

1 Right hemidiaphragm
2 Left hemidiaphragm
3 Liver
4 Stomach
5 Gallbladder
6 Colon (transverse)
7 Colon (ascending)
8 Jejunum
9 Ileum
10 Colon (descending)

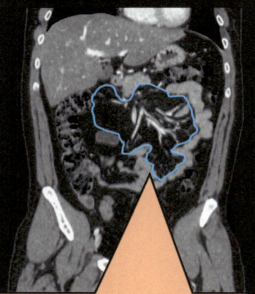

The small bowel mesentery is a broad, fan-shaped fold of peritoneum that suspends the loops of the small intestine from the posterior abdominal wall.

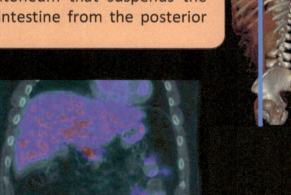

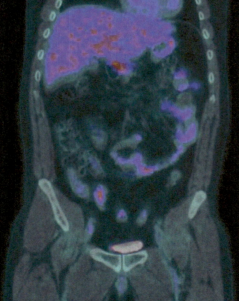

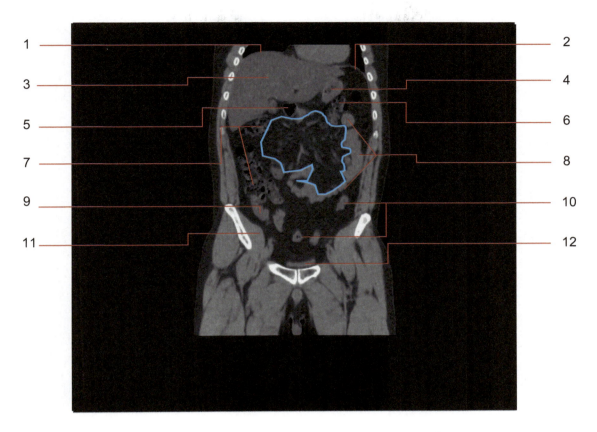

1 Right hemidiaphragm
2 Left hemidiaphragm
3 Liver
4 Stomach
5 Duodenum (bulb)
6 Colon (transverse—left flexure)
7 Colon (ascending)
8 Jejunum
9 Ileum
10 Colon (descending—sigmoid)
11 Right iliopsoas muscle
12 Urinary bladder

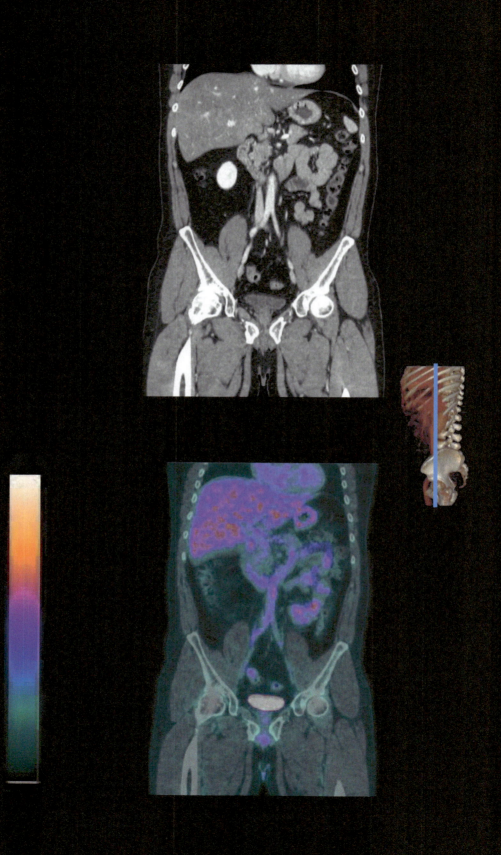

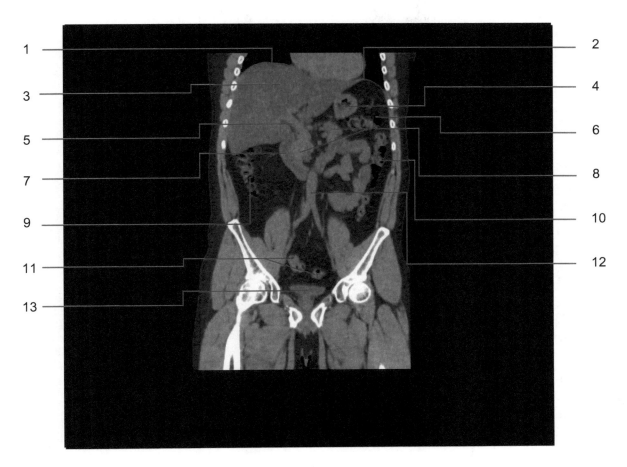

1 Right hemidiaphragm
2 Left hemidiaphragm
3 Liver
4 Stomach
5 Hepatic portal vein
6 Colon (left flexure)
7 Duodenum (descending part)
8 Pancreas
9 Colon (ascending)
10 Colon (descending)
11 Colon (sigmoid)
12 Abdominal aorta
13 Urinary bladder

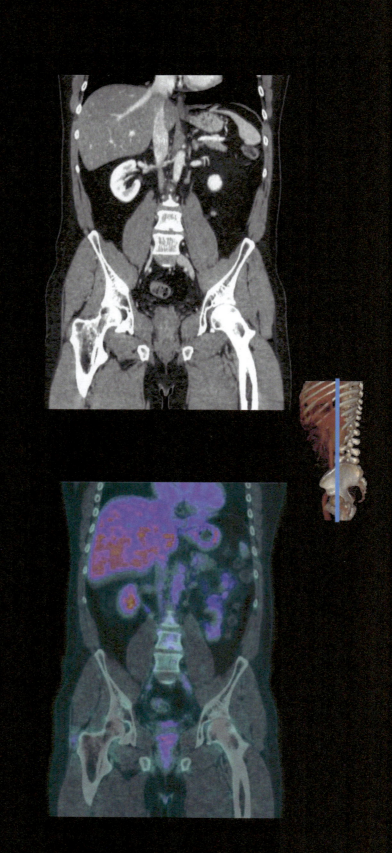

Abdomen and pelvis Chapter | 2 251

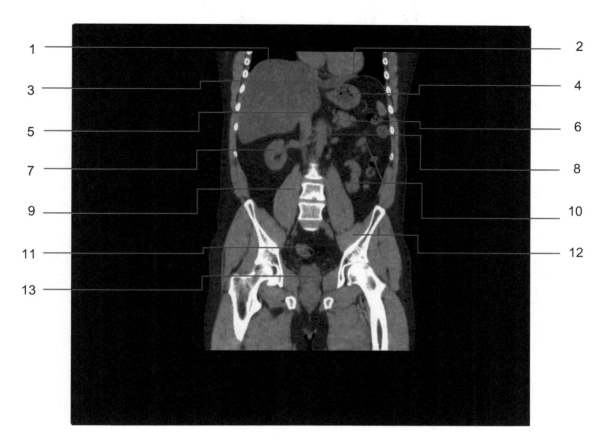

1 Right hemidiaphragm
2 Left hemidiaphragm
3 Liver
4 Stomach
5 Inferior vena cava
6 Pancreas
7 Right kidney
8 Abdominal aorta
9 Right psoas major muscle
10 Jejunum
11 Colon (sigmoid)
12 Left iliacus muscle
13 Urinary bladder

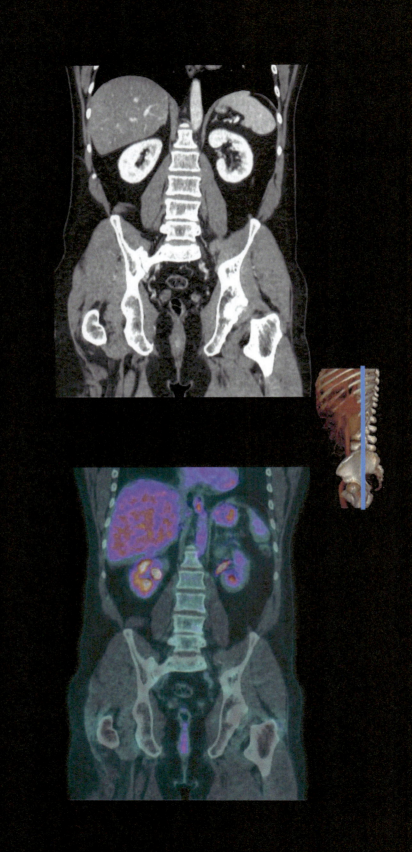

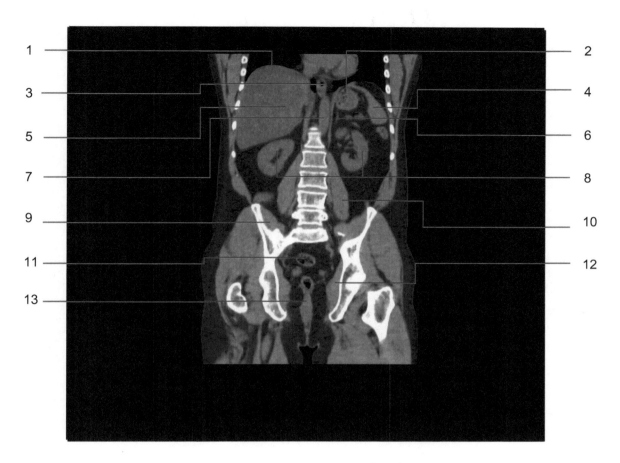

1 Right hemidiaphragm
2 Left hemidiaphragm
3 Esophagus
4 Spleen
5 Liver
6 Left adrenal gland
7 Abdominal aorta
8 Kidneys
9 Right iliacus muscle
10 Left psoas major muscle
11 Colon (sigmoid)
12 Obturator internus muscle
13 Anal canal

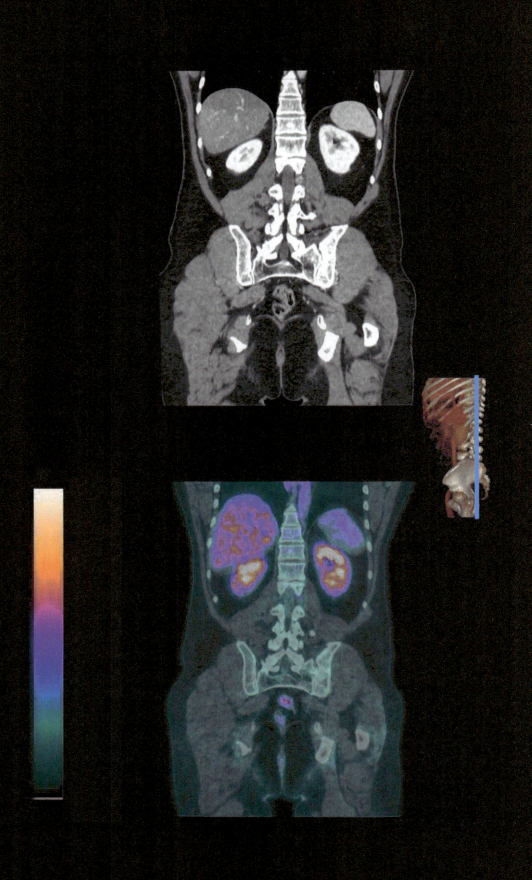

Abdomen and pelvis **Chapter | 2** 255

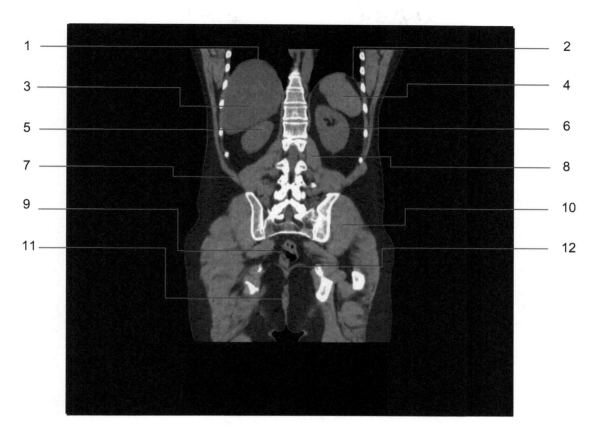

1 Right hemidiaphragm
2 Left hemidiaphragm
3 Liver
4 Spleen
5 Right kidney
6 Left kidney
7 Right latissimus dorsi muscle
8 Left psoas major muscle
9 Rectum
10 Left gluteus muscles
11 Anus
12 Levator ani muscle

2.2 Liver PET/CT
2.2.1 Axial anatomy

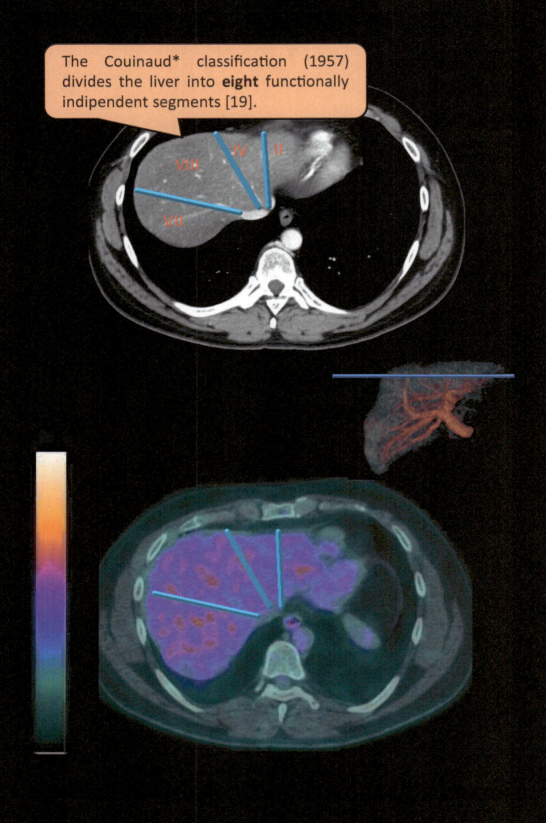

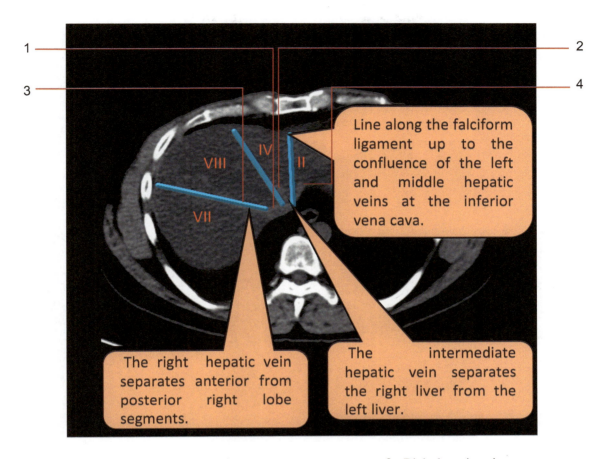

1 Inferior vena cava (retrohepatic)
2 Intermediate (middle) hepatic vein
3 Right hepatic vein
4 Left hepatic vein

CT liver (noncontrast) window width (WW) 200
CT liver (noncontrast) window level (WL) 40

See Ref [19]

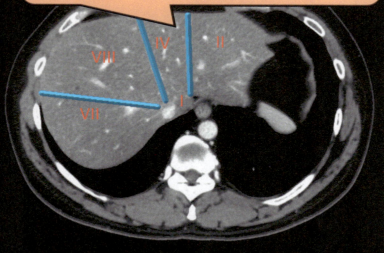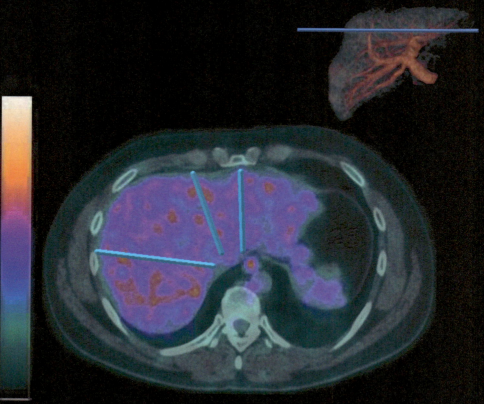

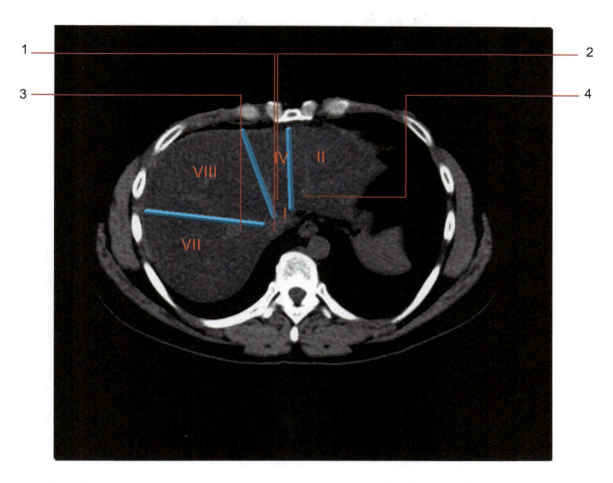

1 Inferior vena cava (retrohepatic)
2 Intermediate (middle) hepatic vein
3 Right hepatic vein
4 Left hepatic vein

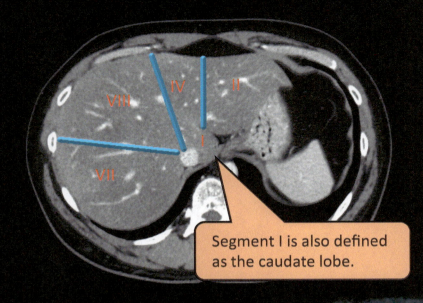
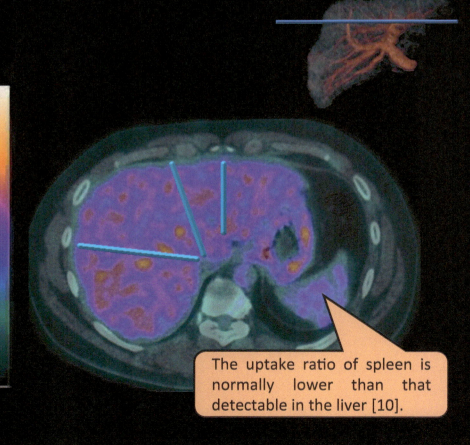

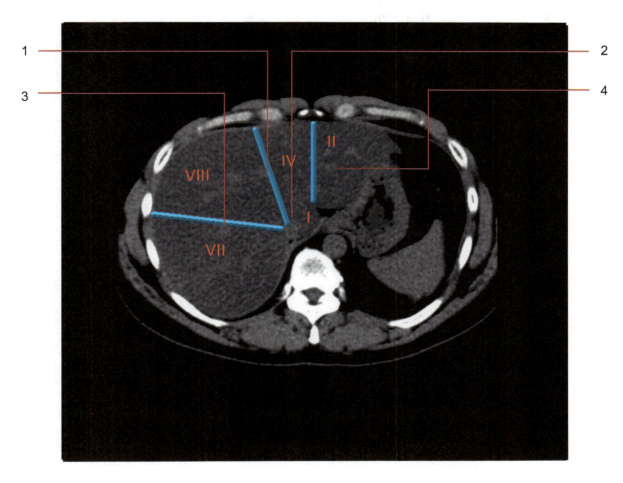

1 Intermediate (middle) hepatic vein
2 Inferior vena cava (retrohepatic)
3 Right hepatic vein
4 Left hepatic vein

See Ref [10]

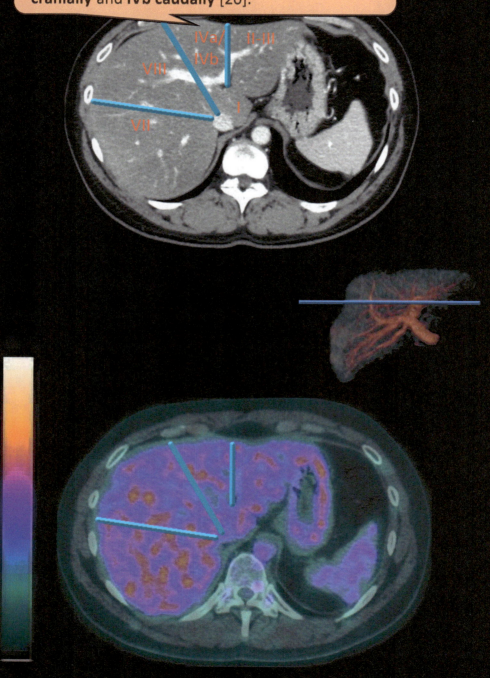

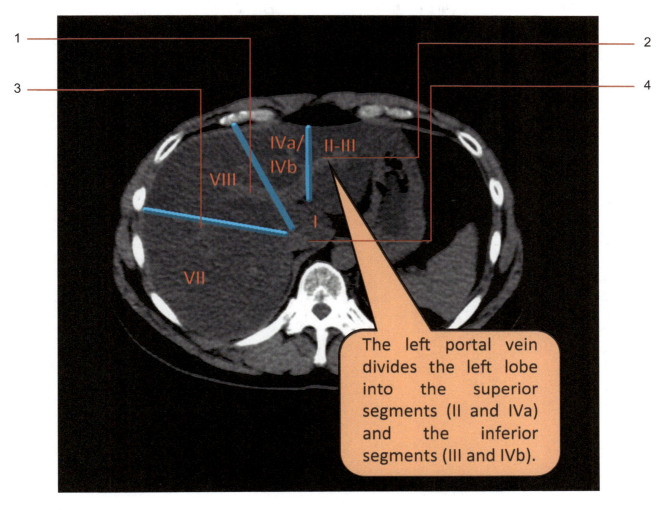

1 Anterior branch from main portal vein (anatomical portal vein variant)
2 Left portal vein
3 Right hepatic vein
4 Inferior vena cava (retrohepatic)

See Ref [20]

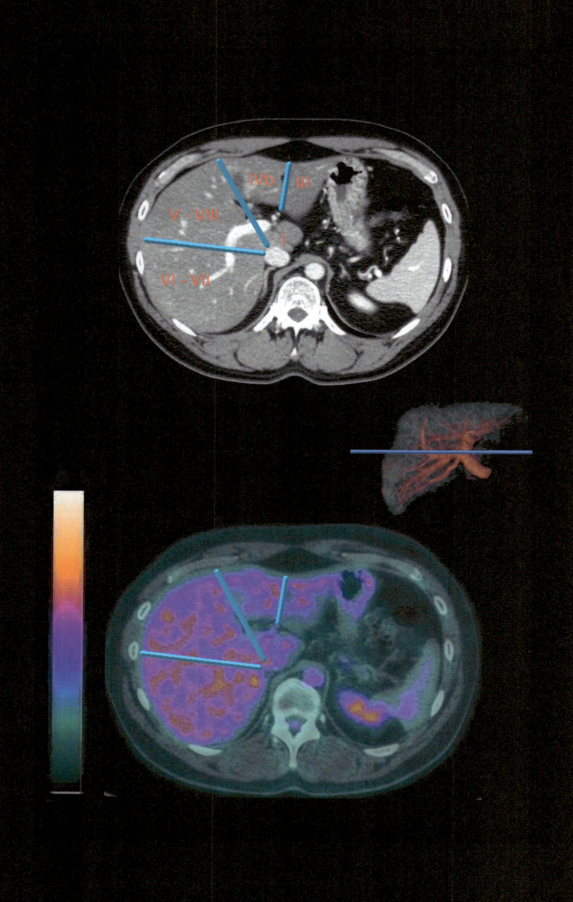

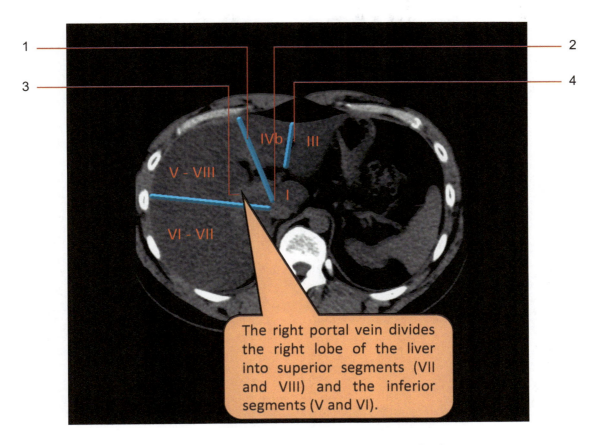

1 Gallbladder fossa and gallbladder
2 Inferior vena cava (retrohepatic)
3 Right portal vein
4 Falciform ligament

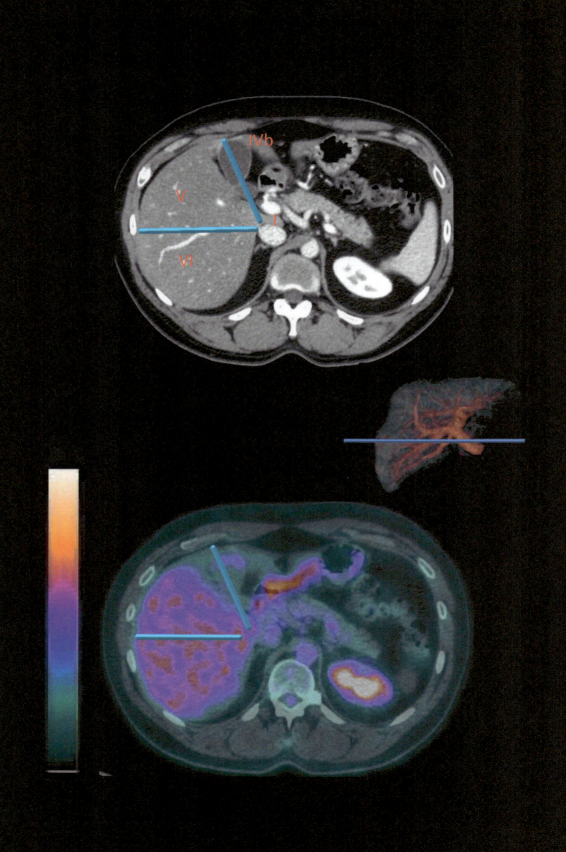

Abdomen and pelvis **Chapter** | 2 267

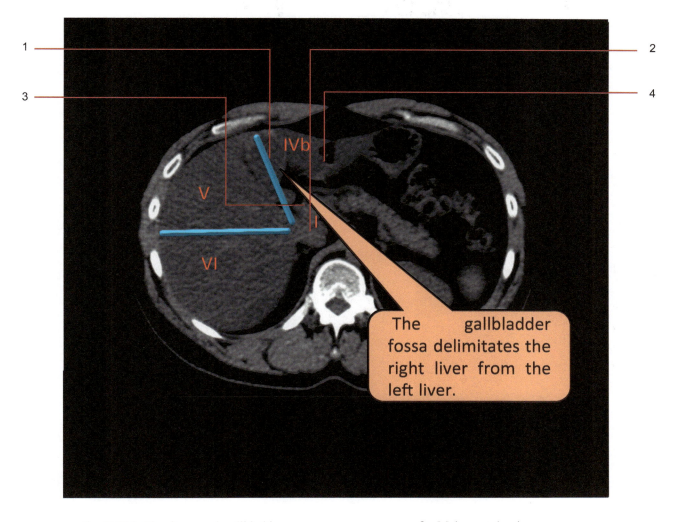

The gallbladder fossa delimitates the right liver from the left liver.

1 Gallbladder fossa and gallbladder
2 Inferior vena cava (retrohepatic)
3 Main portal vein
4 Falciform ligament

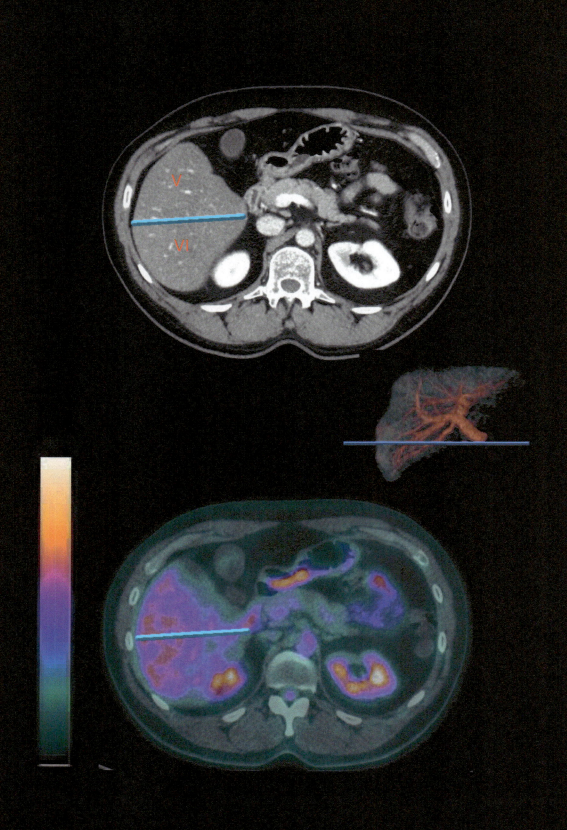

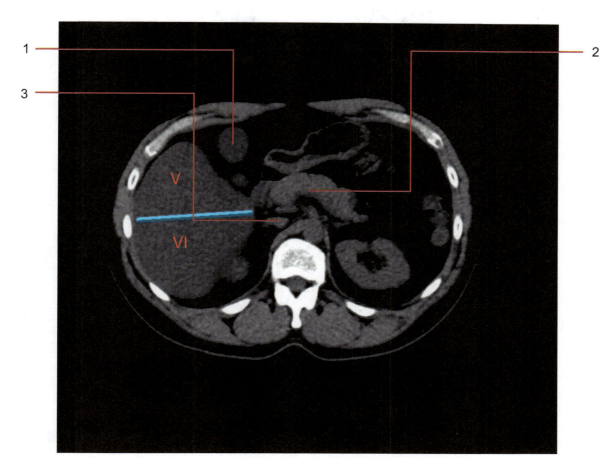

1 Gallbladder
2 Splenic vein
3 Inferior vena cava (infrahepatic)

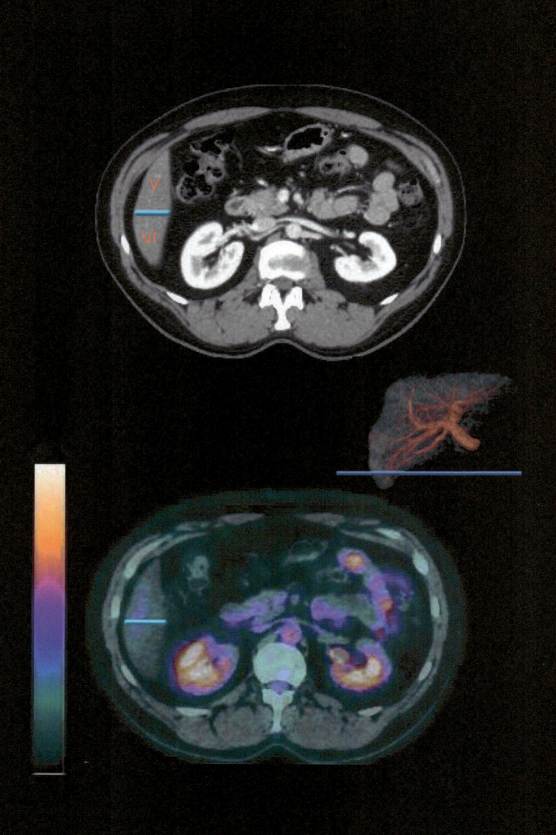

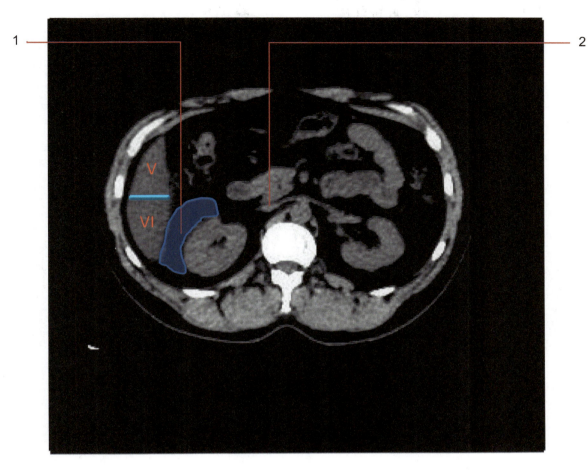

1 Hepatorenal recess (subhepatic recess) **2** Inferior vena cava (infrahepatic)

2.3 Peritoneum and retroperitoneum PET/CT
2.3.1 Peritoneum

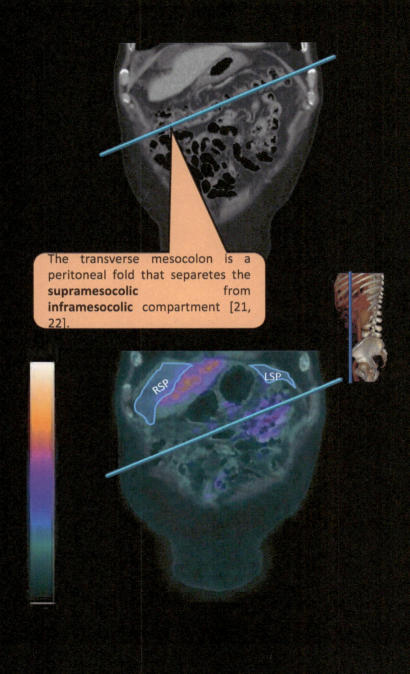

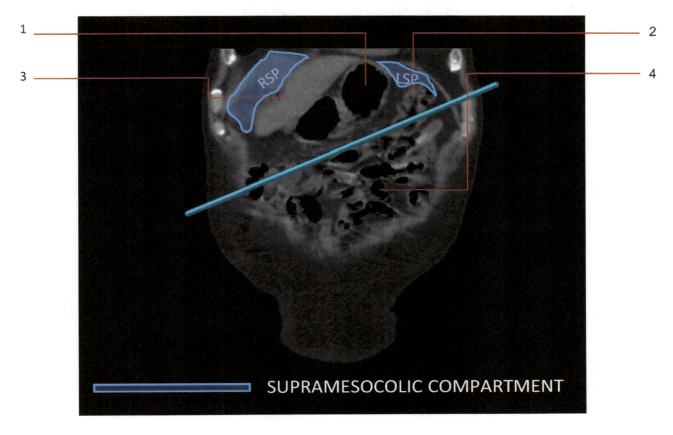

1 Stomach
2 Colon
3 Liver
4 Jejunum

Supramesocolic peritoneal spaces:
Right subphrenic space (RSP)
Left subphrenic space (LSP)
Right subhepatic space or Morison's pouch or hepatorenal space (RSHS)
Left subhepatic space (LSHS)
Perisplenic space (PSS)
Lesser sac or omental bursa (LS)
See Refs [21,22]

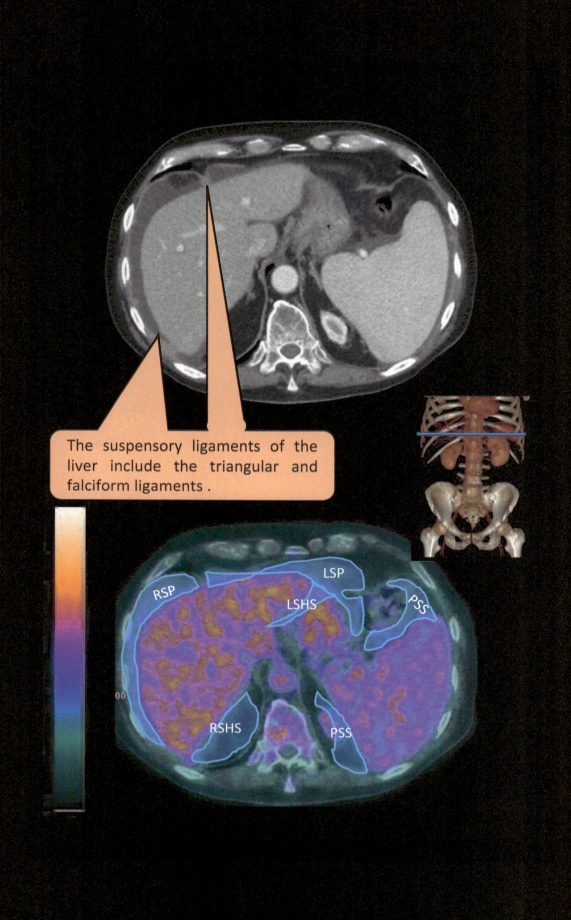

Abdomen and pelvis Chapter | 2   275

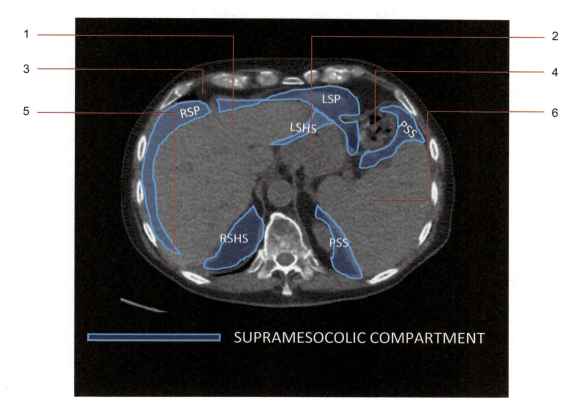

1 Liver
2 Stomach
3 Falciform ligament
4 Colon (left flexure)
5 Triangular ligament (thickened)
6 Spleen

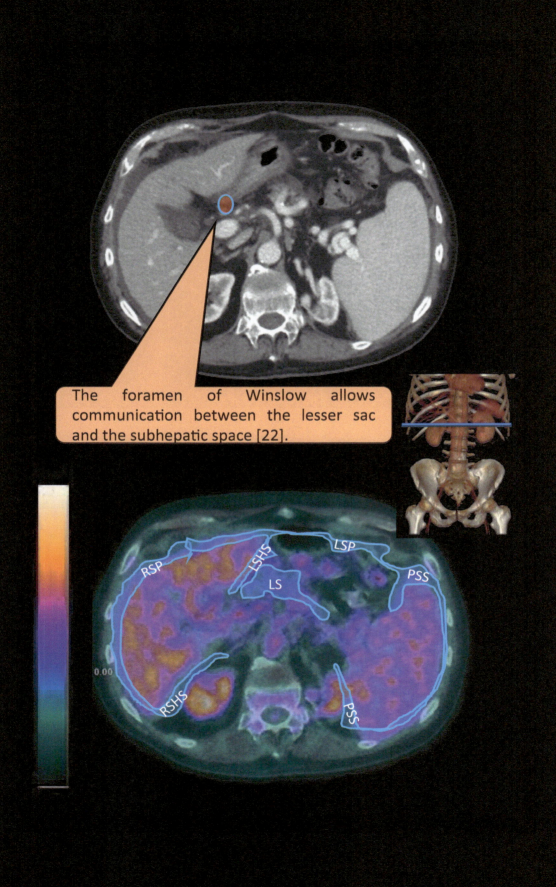

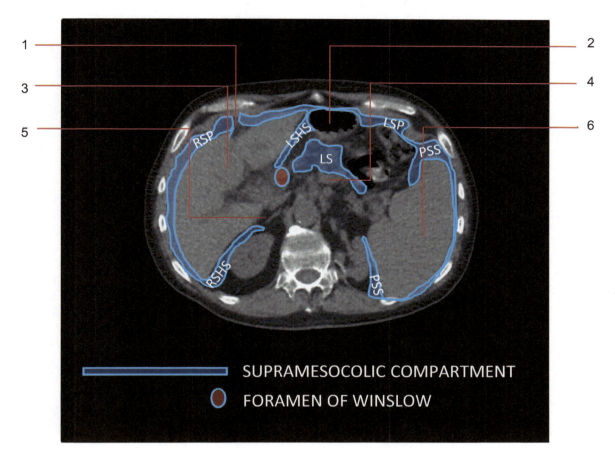

1 Falciform ligament
2 Stomach
3 Liver
4 Pancreas
5 Right adrenal gland
6 Spleen

See Ref [22]

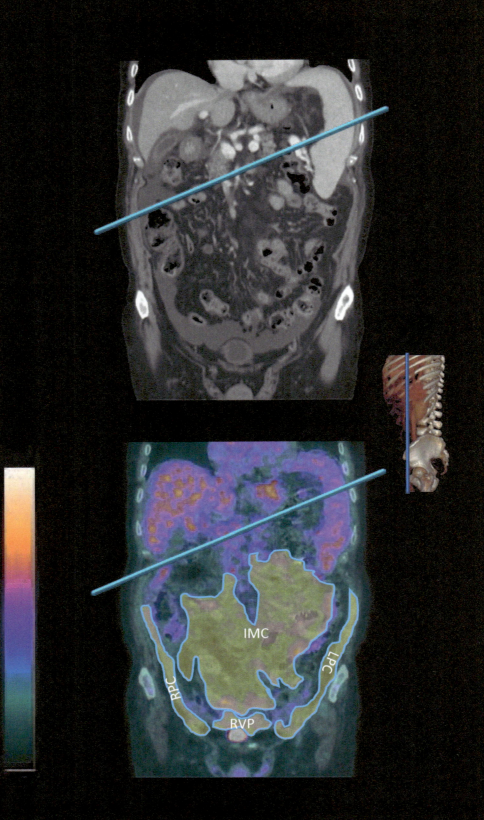

Abdomen and pelvis Chapter | 2   279

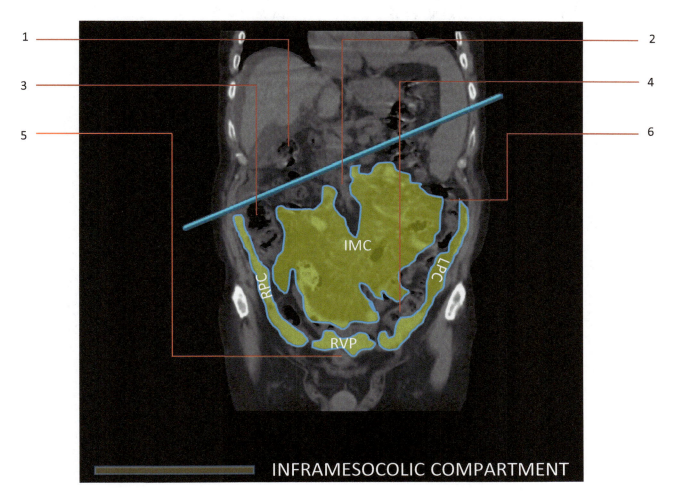

1 Colon (right flexure)
2 Root of mesentery
3 Colon (ascending)
4 Colon (sigmoid)
5 Bladder
6 Colon (descending)

Inframesocolic Peritoneal Spaces:
Inframesocolic space (IMC)
Right paracolic space or gutter (RPC)
Left paracolic space or gutter (LPC)
Rectovesical pouch in men or rectouterine/Douglas pouch in female (RVP)

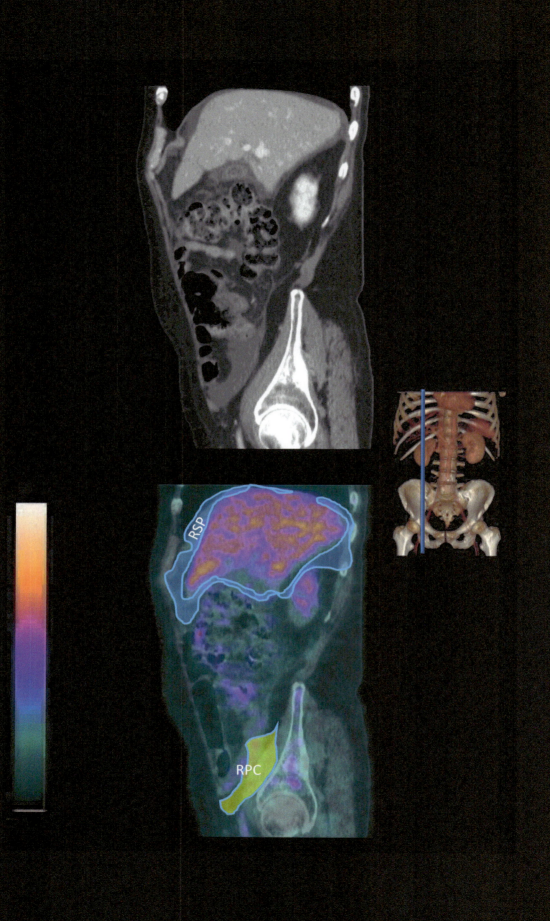

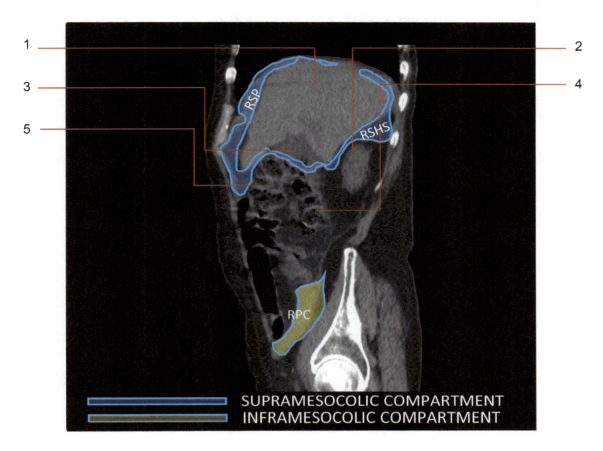

1 Liver
2 Right kidney
3 Gallbladder
4 Colon (ascending)
5 Colon (right flexure)

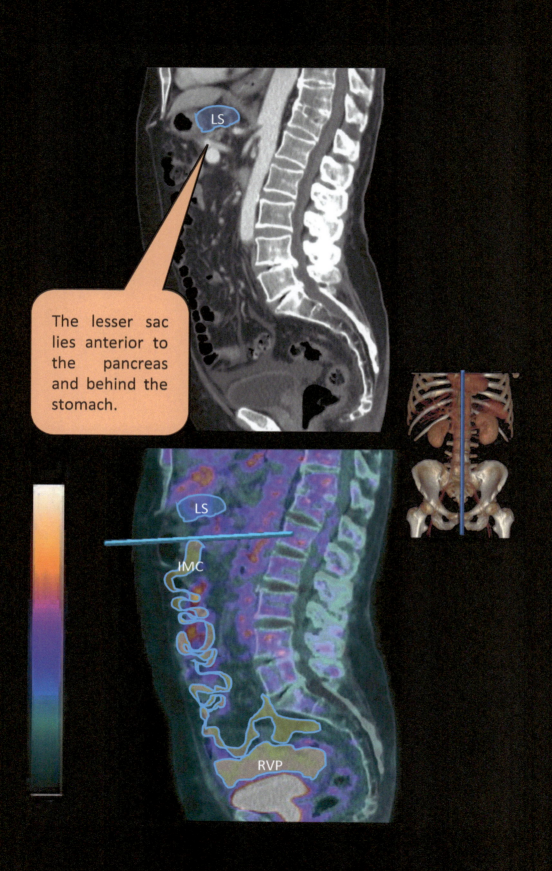

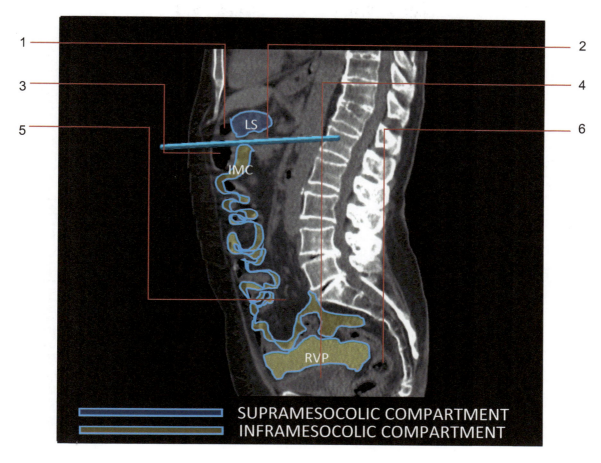

1 Stomach
2 Pancreas (body)
3 Colon (transverse)
4 Urinary bladder
5 Mesentery
6 Rectum

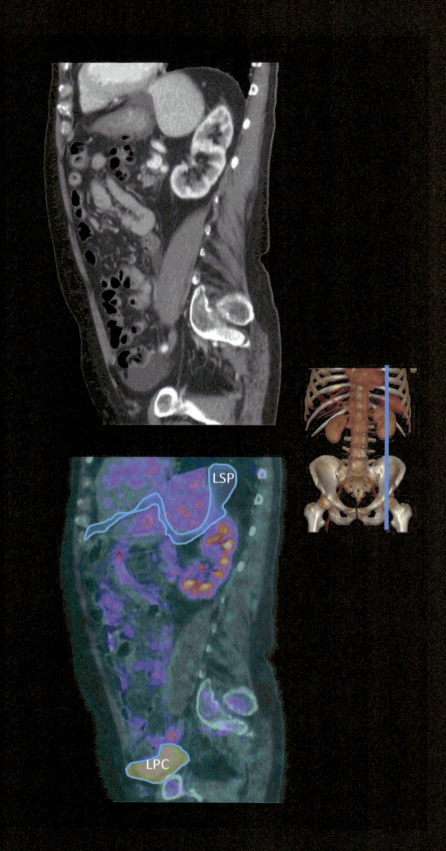

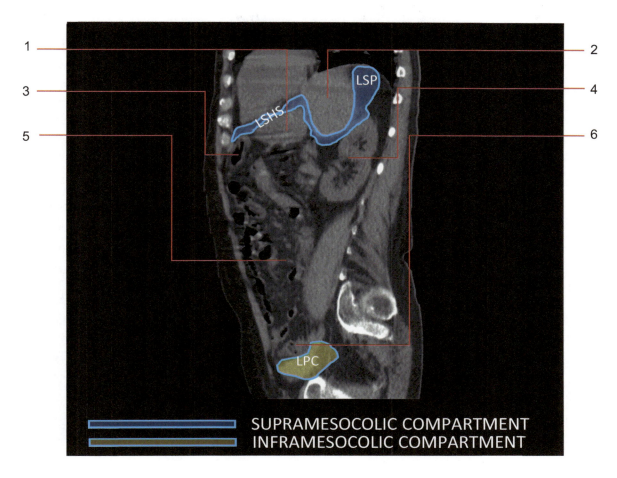

1 Stomach
2 Spleen
3 Colon (transverse)
4 Left kidney
5 Mesentery
6 Colon (sigmoid)

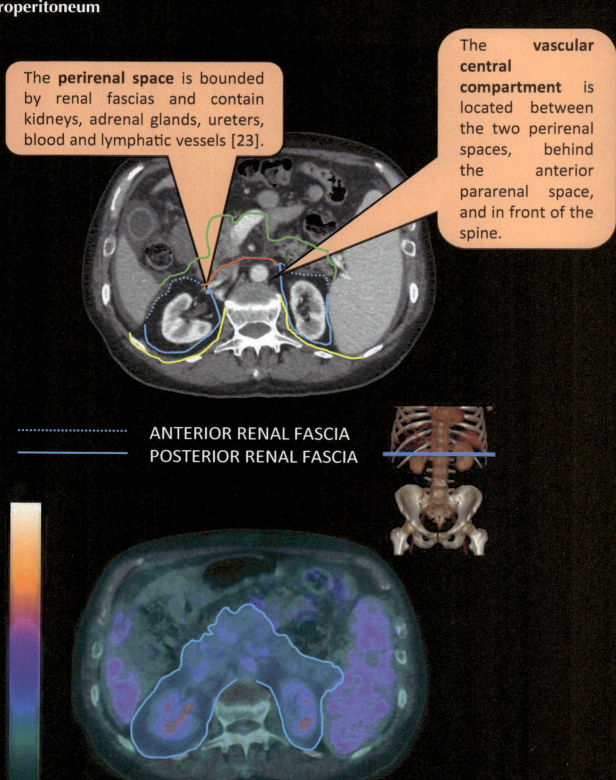

Abdomen and pelvis **Chapter** | **2**   287

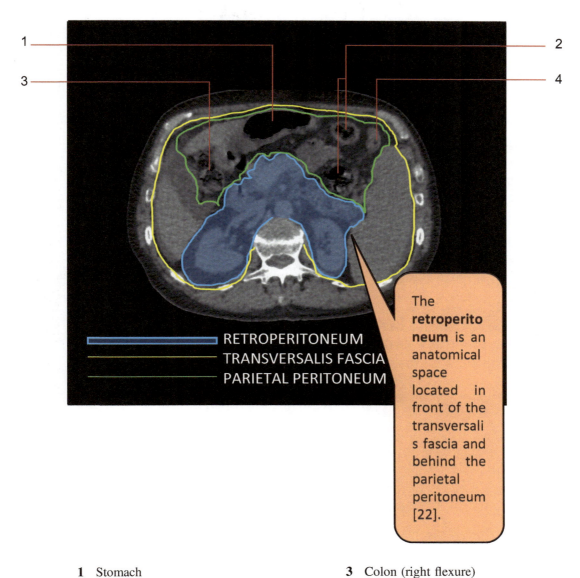

RETROPERITONEUM
TRANSVERSALIS FASCIA
PARIETAL PERITONEUM

The **retroperitoneum** is an anatomical space located in front of the transversalis fascia and behind the parietal peritoneum [22].

1 Stomach
2 Colon (left flexure)
3 Colon (right flexure)
4 Jejunum

See Refs [22,23]

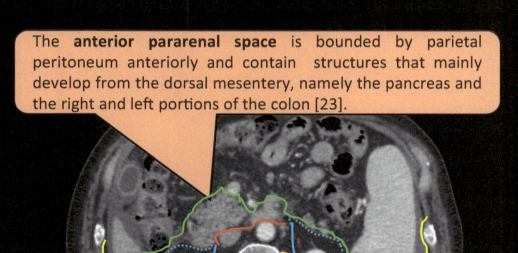

The **anterior pararenal space** is bounded by parietal peritoneum anteriorly and contain structures that mainly develop from the dorsal mesentery, namely the pancreas and the right and left portions of the colon [23].

The **posterior pararenal space** is bounded by transversalis fascia posteriorly and contains adipose tissue [22].

.... ANTERIOR RENAL FASCIA
———————— POSTERIOR RENAL FASCIA

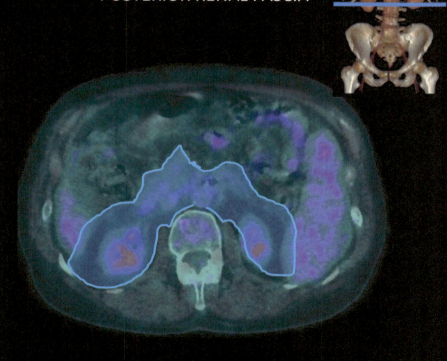

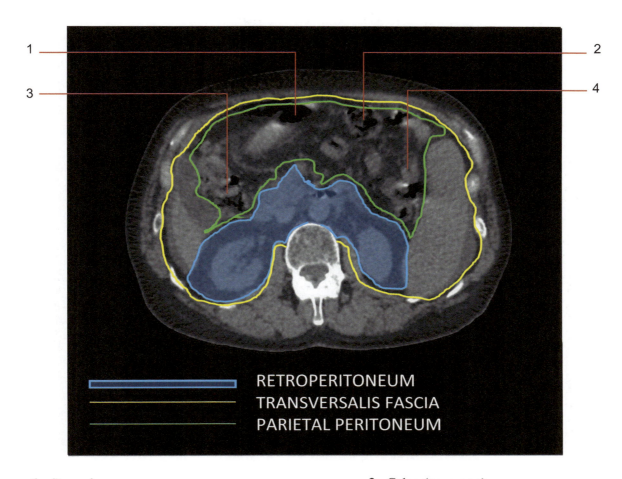

1 Stomach
2 Colon (left flexure)
3 Colon (transverse)
4 Jejunum

See Refs [22,23]

The **perirenal space** usually, but not always, cut off inferiorly by the fusion of renal fascias and does not extend into the pelvis [23].

The **posterior pararenal space** is bounded by transversalis fascia posteriorly and contains adipose tissue [22].

POSTERIOR RENAL FASCIA
LATEROCONAL FASCIA

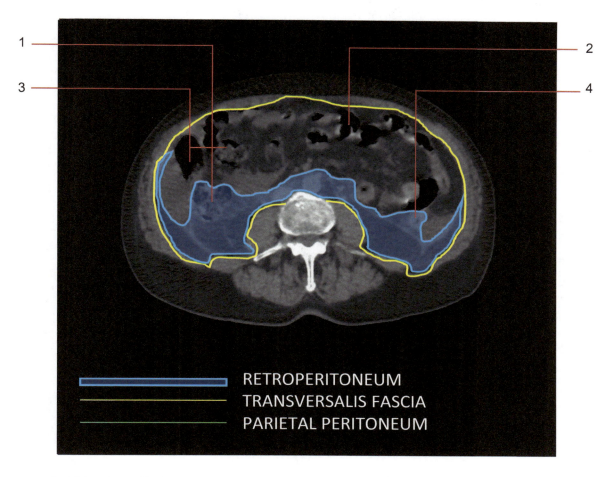

1 Colon (ascending)
2 Jejunum
3 Colon (right flexure)
4 Colon (descending)

See Refs [22,23]

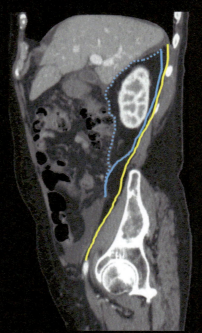
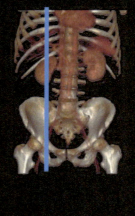

.... ANTERIOR RENAL FASCIA
——— POSTERIOR RENAL FASCIA
——— TRANSVERSALIS FASCIA

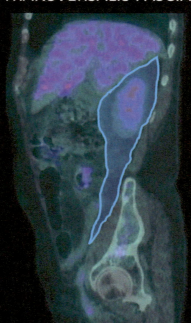

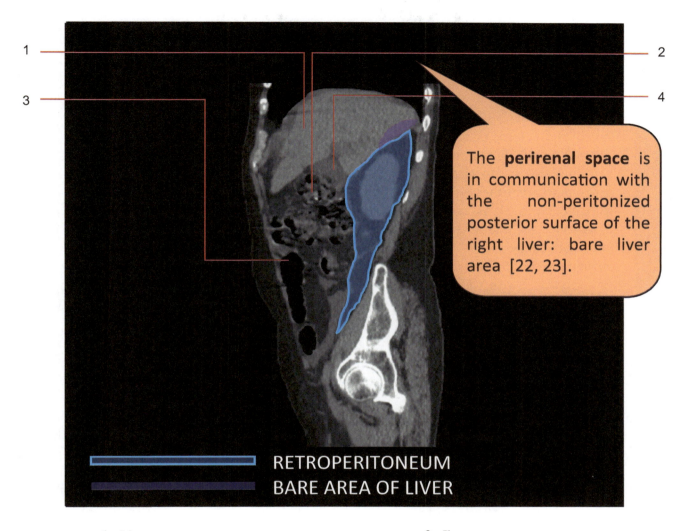

1 Liver
2 Colon (right flexure)
3 Ileum
4 Gallbladder

See Refs [22,23]

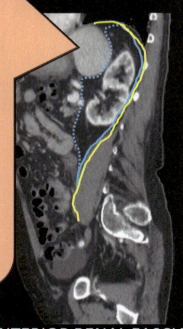

> ^{18}F-FDG PET allows detection of small malignant nodes not identified or not meeting size criteria for malignancy with computed tomography (CT) and tumor recurrences in surgical beds that are otherwise difficult to assess. However, evaluation of retroperitoneal malignancies or adenopathy with FDG PET can be complicated by urinary and colonic activity or anatomic variants [24].

·········· ANTERIOR RENAL FASCIA
────── POSTERIOR RENAL FASCIA
────── TRANSVERSALIS FASCIA

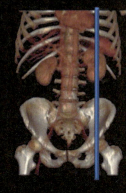

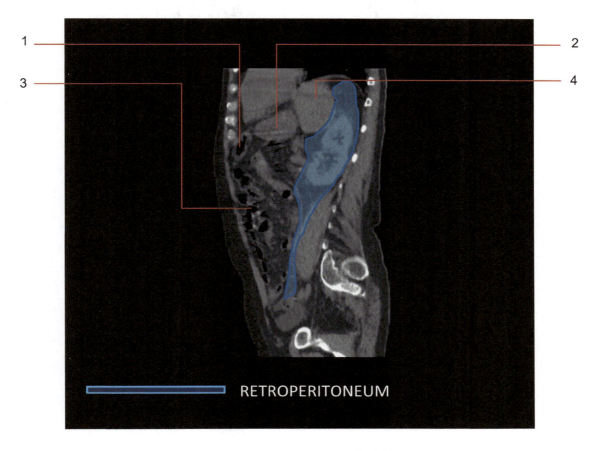

1 Colon (transverse)
2 Stomach
3 Jejunum
4 Spleen

See Ref [24]

2.4 Pelvis and perineum PET/MRI
2.4.1 Female pelvis

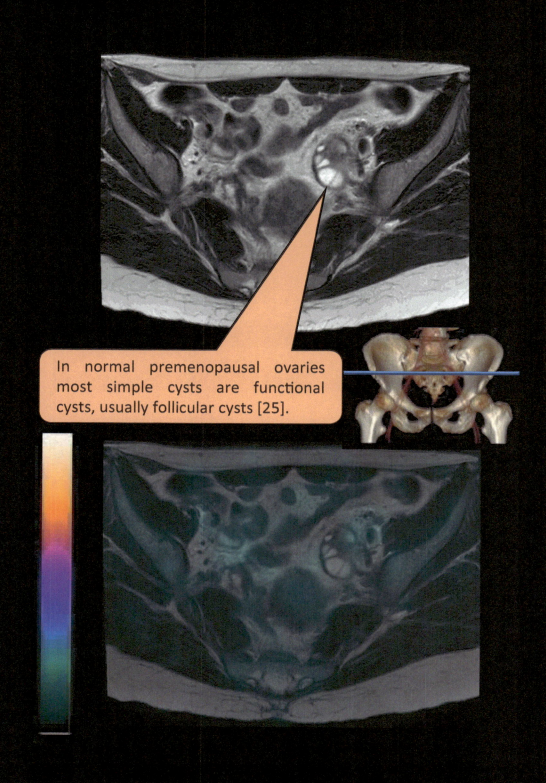

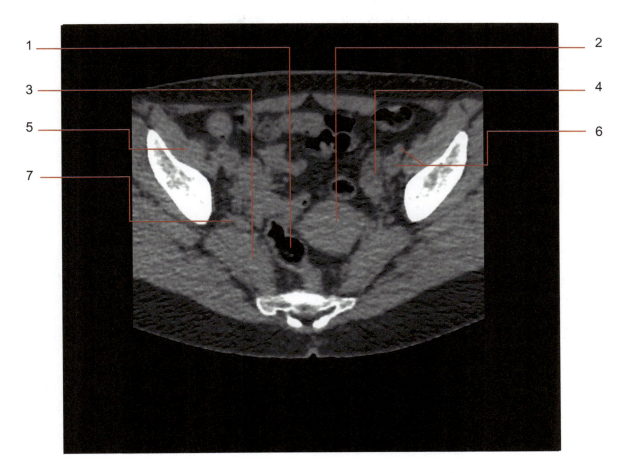

1 Left ovary
2 Uterus (fundus)
3 Right piriformis muscle
4 Sigmoid colon
5 Right iliacus muscle
6 Left external iliac artery (in front) and vein (in back)
7 Right internal artery and vein

See Ref [25]

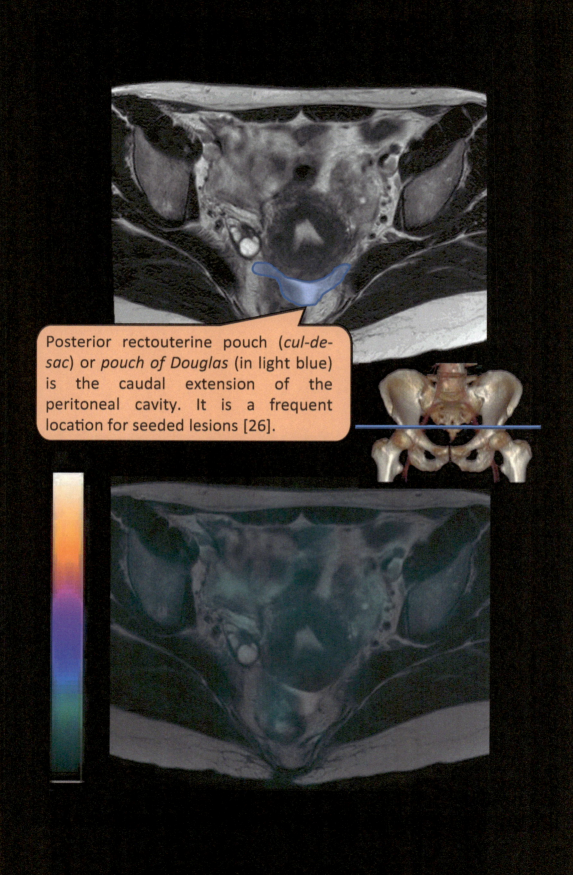

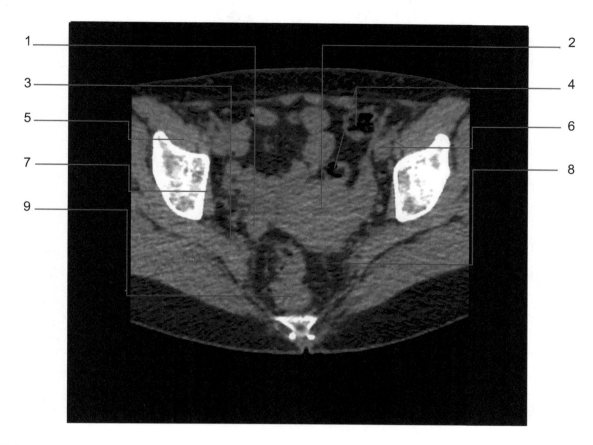

1 Right ovary
2 Uterus (body and uterine cavity)
3 Right piriformis muscle
4 Descending and sigmoid colon
5 Right iliacus muscle

6 Left external iliac artery (in front) and vein (in back)
7 Right obturator internus muscle
8 Rectouterine pouch (of Douglas)
9 Rectum

See Ref [26]

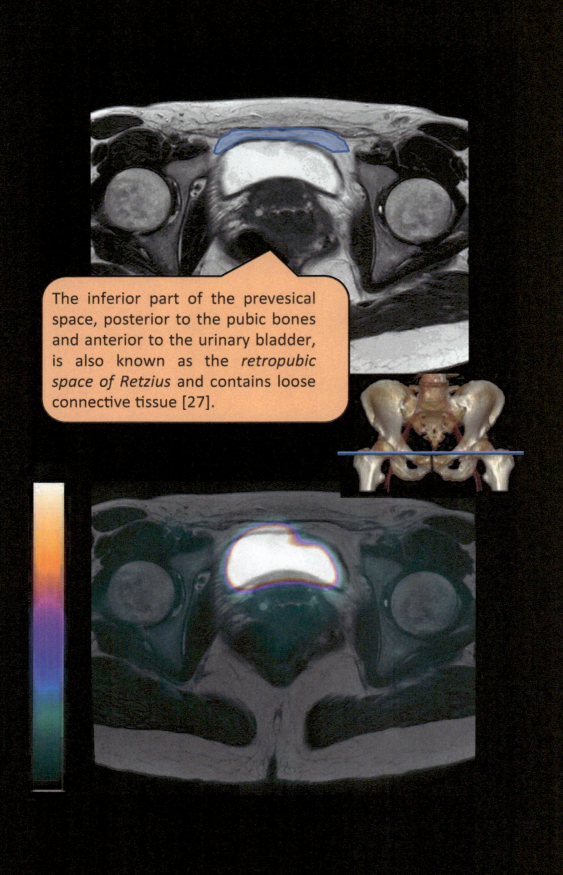

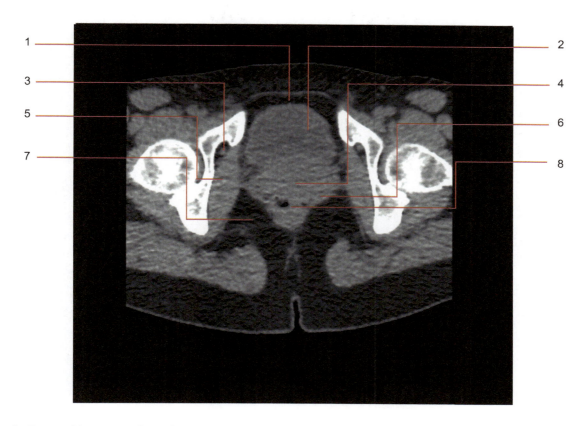

1 Retropubic space (of Retzius)
2 Urinary bladder
3 Right obturator canal
4 Cervix of uterus
5 Right obturator internus muscle
6 Left puborectalis muscle
7 Right ischiorectal fossa
8 Rectum

See Ref [27]

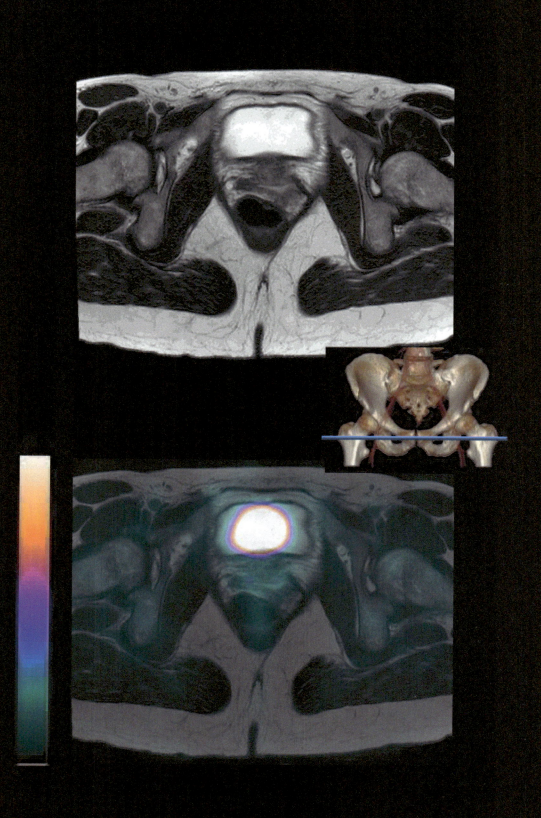

Abdomen and pelvis Chapter | 2   303

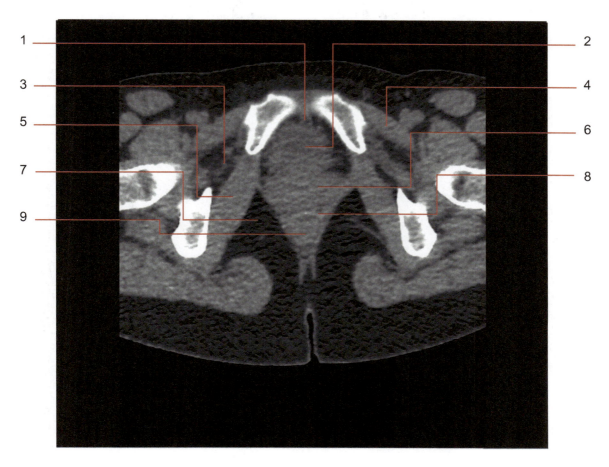

1 Retropubic space (of Retzius)
2 Urinary bladder
3 Right obturator canal
4 Left pectineus muscle
5 Right obturator internus muscle
6 Vagina
7 Right ischioanal fossa
8 Left puborectalis muscle
9 Anal canal

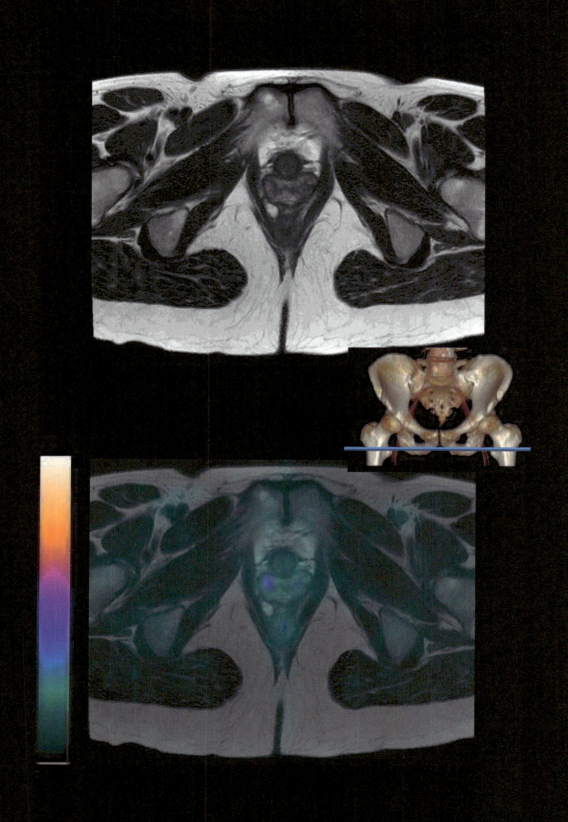

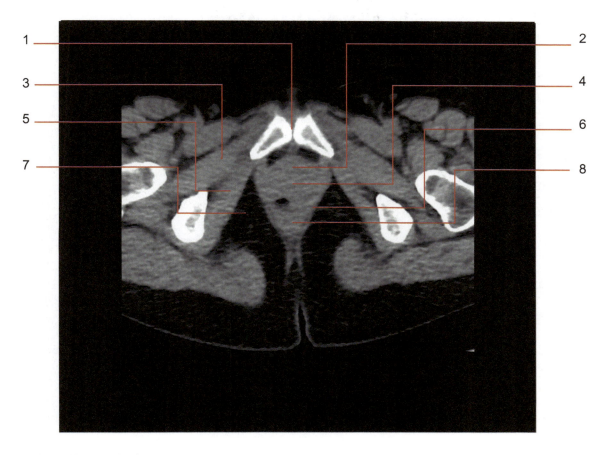

1 Pubic symphysis
2 Urethra
3 Right obturator externus muscle
4 Vagina
5 Right obturator internus muscle
6 Left puborectalis muscle
7 Right ischioanal fossa
8 Anal canal

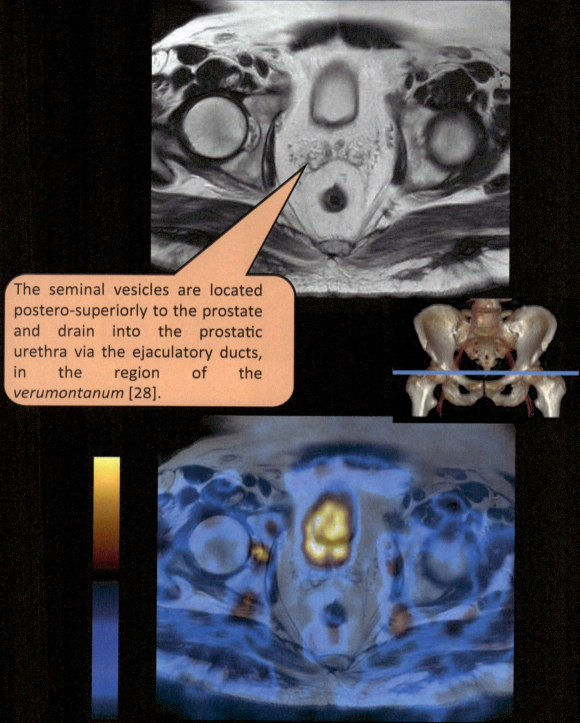

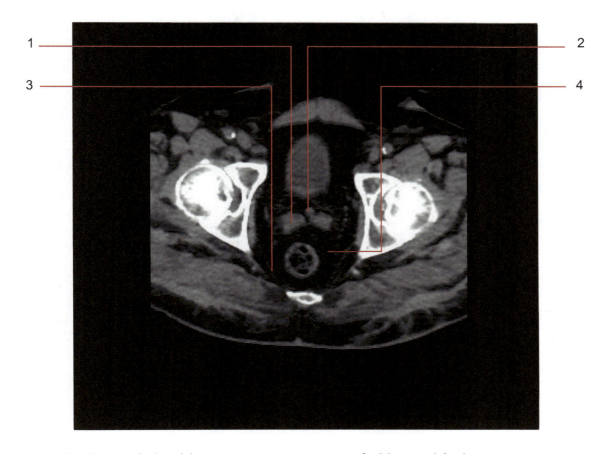

1 Right seminal vesicle
2 Left ductus deferens
3 Mesorectal fascia
4 Perirectal space

See Ref [28]

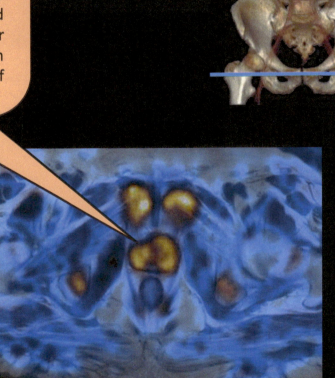

^{11}C- or ^{18}F-choline PET/CT has limited role in the detection of primary prostate cancer because of limited sensitivity and specificity for the differentiation of benign from malignant prostatic lesions. Its role may be valid in high risk prostate cancer patients with high PSA serum level, in the detection of secondary lesions [29].

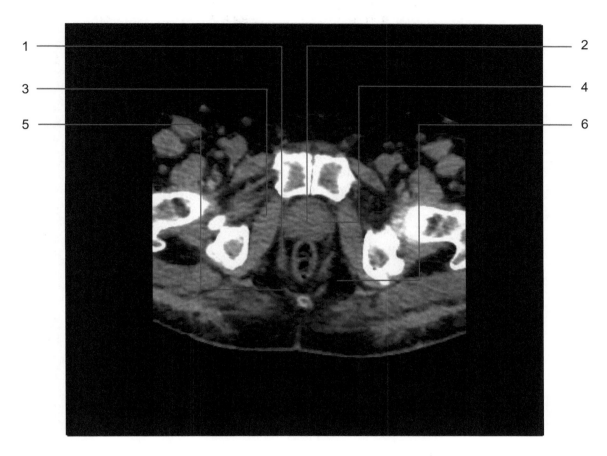

1 Right puborectalis muscle
2 Prostate gland ("central gland")
3 Right obturator internus muscle
4 Prostate gland (peripheral zone)
5 Right pubococcygeus muscle
6 Left ischiorectal fossa

See Ref [29]

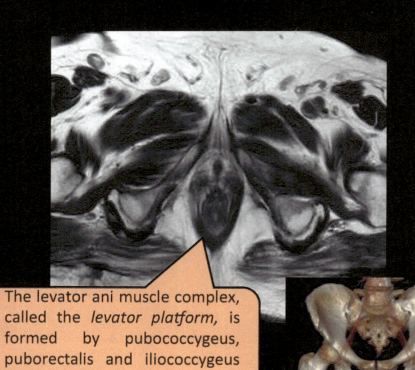

The levator ani muscle complex, called the *levator platform*, is formed by pubococcygeus, puborectalis and iliococcygeus muscles.

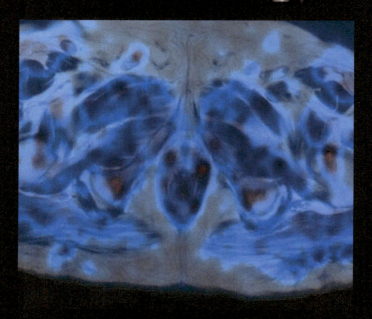

Abdomen and pelvis Chapter | 2 311

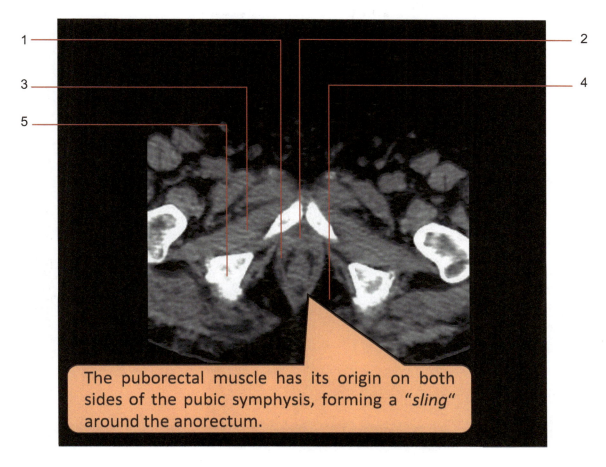

The puborectal muscle has its origin on both sides of the pubic symphysis, forming a *"sling"* around the anorectum.

1 Right puborectalis muscle
2 Urethra
3 Right obturator internus muscle
4 Left ischiorectal fossa
5 Right ischial tuberosity

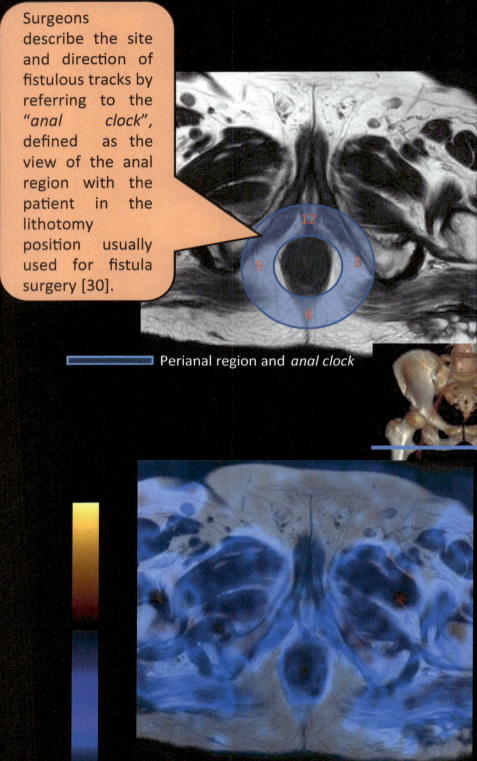

Perianal region and *anal clock*

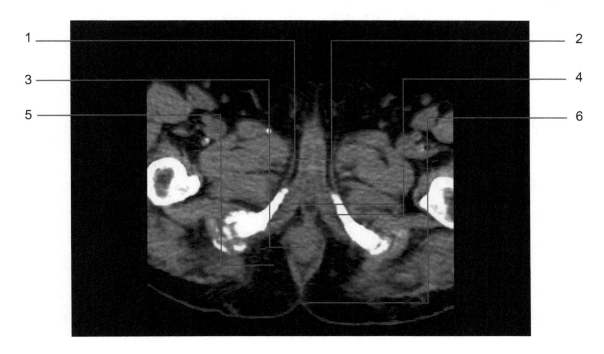

1 Right corpus cavernosum
2 Left corpus cavernosum
3 Anus
4 Right and left ischiocavernosus muscles
5 Right ischioanal fossa
6 Intergluteal cleft

See Ref [30]

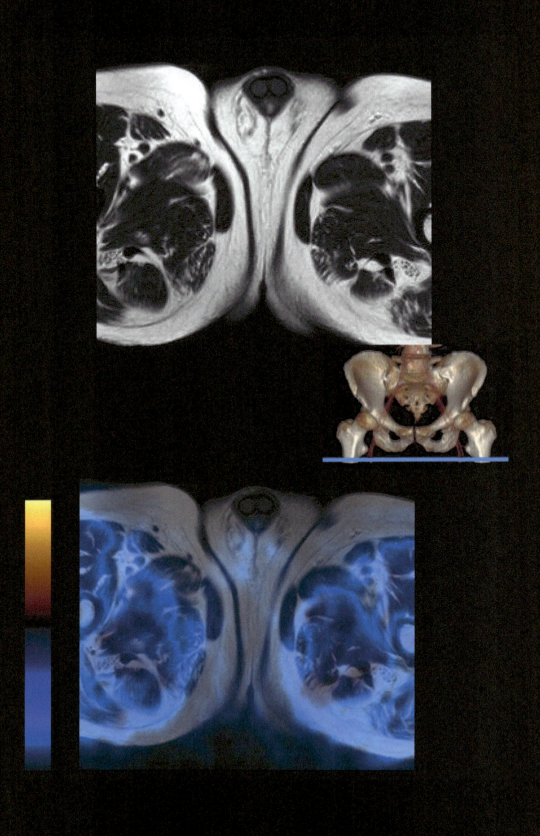

Abdomen and pelvis Chapter | 2   315

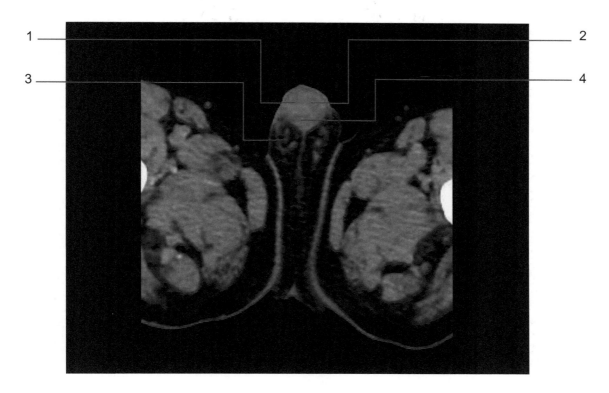

1   Right corpus cavernosum
2   Left corpus cavernosum
3   Right pampiniform plexus and ductus deferens
4   Corpus spongiosum

2.5 Clinical cases, tricks, and pitfalls
2.5.1 ^{18}F-FDG

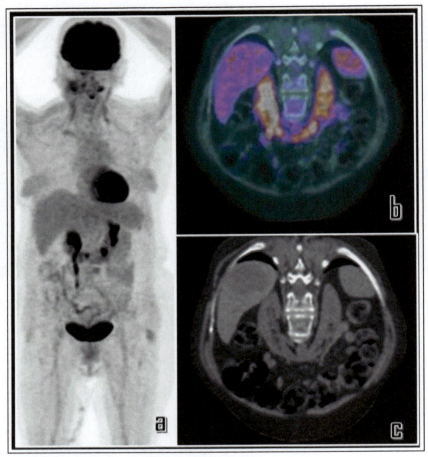

FIGURE 2.1 A 66-year-old female patient underwent ^{18}F-FDG PET/CT during follow-up of lung cancer. No pathologic findings were detected at whole body PET (a). Oblique PET/CT (b) and CT (c) views eloquently show a condition of horseshoe kidney, an uncommon congenital disorder usually diagnosed incidentally, as in this case, where both kidneys are fused together to form a horseshoe-shape or a single *ren arcuatus*.

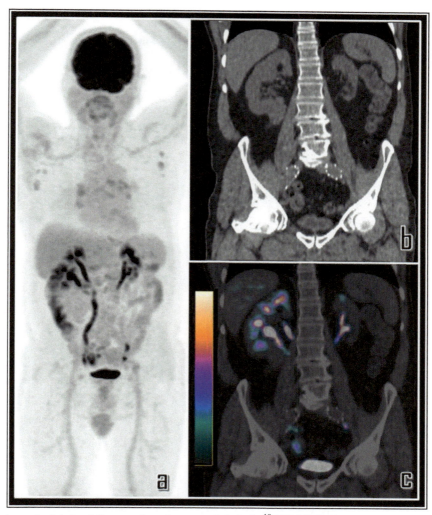

FIGURE 2.2 In a patient examined during follow-up of rectal cancer, whole body ^{18}F-FDG PET *Maximum Intensity Projection* (a) does not show pathologic tracer uptake. Conversely, a condition of bilateral ureteral duplex collecting system is visible and confirmed in correlative coronal CT (b) and PET/CT views (c). Particularly, in this patient, we observed a condition of bifid ureter: two ureters that unite before emptying into the bladder, bilaterally.

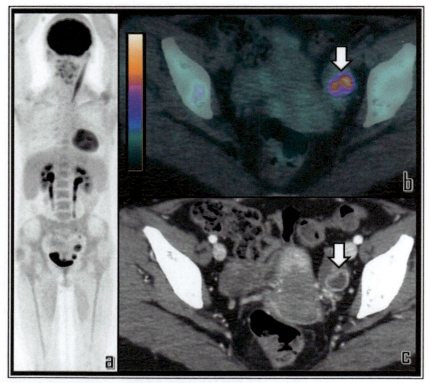

FIGURE 2.3 A 39-year-old woman was examined by whole body ^{18}F-FDG PET/CT, during follow-up of mediastinal Hodgkin's lymphoma. In PET, *Maximum Intensity Projection* (a) is detectable focal uptake in left pelvis, corresponding to a hypodense lesion in axial PET/CT (b), more evident as cystic, strongly enhancing periphery at correlative axial contrast-enhanced CT (c), consensually performed. This finding was identified as *corpus luteum* [31]. No foci of abnormal tracer uptake were detected at whole body scan.

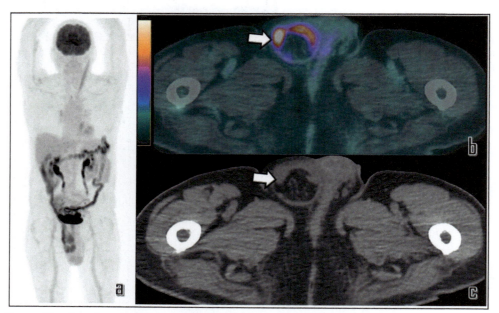

FIGURE 2.4 A 55-year-old male patient, previously submitted to radiotherapy on mediastinal *non-Hodgkin's lymphoma* and bone marrow autotransplant, underwent ^{18}F-FDG PET/CT during follow-up. Whole body PET *Maximum Intensity Projection* (a) was negative for disease relapse. Axial PET/CT detail of the pelvis (b, *arrow*) shows intense tracer uptake in right inguinal hernia within the scrotum, also evident in correlative axial CT view (c, *arrow*).

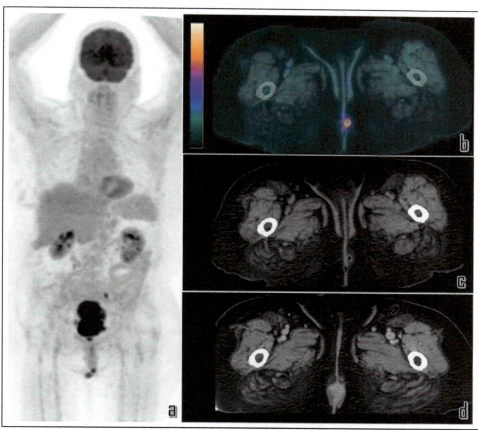

FIGURE 2.5 A 45-year-old patient was examined by whole body ^{18}F-FDG PET/CT during follow-up of gastrointestinal stromal tumor (GIST). PET *Maximum Intensity Projection* (a) was negative. An area of focal uptake was observed in left perineum, more evident in axial PET/CT (b) and low-dose CT (c) views, with air−fluid collection in the left perianal soft tissue and mild contrast enhancement correlative axial contrast-enhanced CT (d). Morphologic features allowed diagnosis of perianal fistula; active flogosis was evidenced by PET [32].

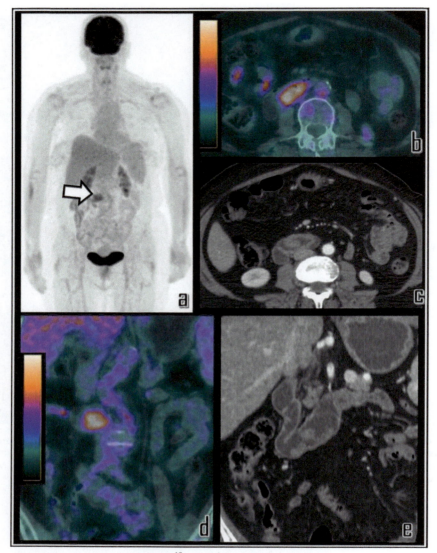

FIGURE 2.6 A 61-year-old female patient was examined by ^{18}F-FDG PET/CT during follow-up of *non-Hodgkin's lymphoma*. Whole body PET *Maximum Intensity Projection* (a, arrow) only showed single area of moderate tracer uptake in the abdomen, corresponding to parietal thickening of the duodenum with mild, homogenous contrast enhancement in correlative diagnostic CT (c). This finding is also showed in coronal PET/CT (d) and CT (e) views. Patient underwent biopsy. Histological exam diagnosed duodenal tubulovillous adenoma.

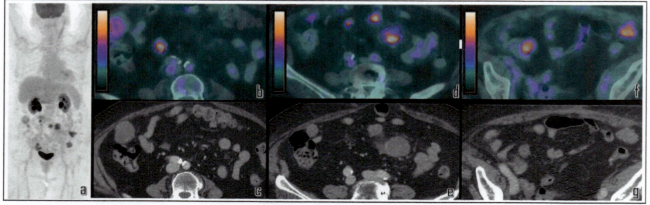

FIGURE 2.7 A 59-year-old patient, previously submitted to bilateral oophorectomy and hysterectomy for ovarian cancer, underwent contrast-enhanced CT and ^{18}F-FDG PET/CT in single session, due to rise of Ca125 (*136 U/mL at the time of the scan*). Whole body PET (a) showed several areas of pathologic tracer uptake in the abdomen, corresponding to mesenteric hypodense nodes in axial PET/CT views with intense contrast enhancement in correlative CT views (b–g). A diagnosis of peritoneal carcinomatosis was done. In ovarian cancer patients, high Ca125 serum levels may be predictor of positive results at ^{18}F-FDG PET/CT [33].

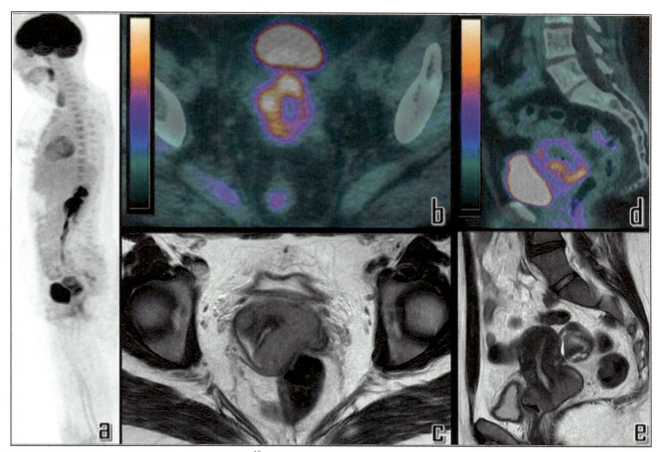

FIGURE 2.8 A 55-year-old patient was examined by ^{18}F-FDG PET/CT for staging uterine cervical cancer. Whole body PET *Maximum Intensity Projection*, in left lateral view, (a) underlined pathologic ^{18}F-FDG uptake in the cervix, as evident in axial PET/CT view (b), in correspondence to a uterine lesion of the cervix, with parametrial invasion, showed in axial T2-weighted MRI, subsequently performed (c). Sagittal PET/CT (d) and T2-weighted MRI (e) views also permit to assess concurrent infiltration of vaginal fornix. No other foci of pathologic tracer uptake were observed at whole body PET/CT scan.

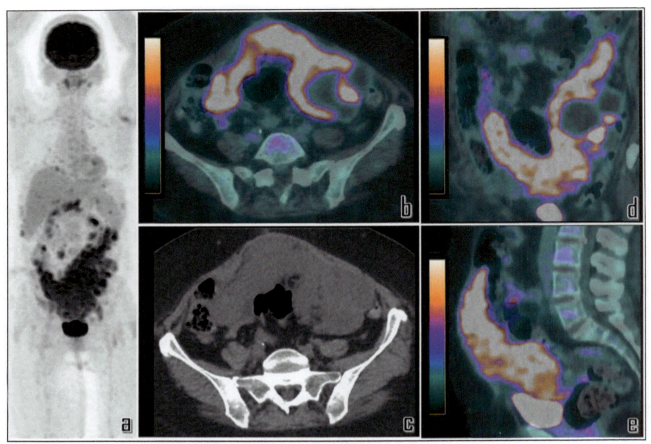

FIGURE 2.9 62-year-old woman with history of treated ovarian cancer, for which she had undergone surgery and adjuvant chemo- and radiotherapy. Peritoneal recurrence occurred 3 years later, which was treated with chemotherapy, achieving apparent complete remission. One year later, the patient presented with lower abdominal pain and increasing abdominal distention and underwent whole body ^{18}F-FDG PET/CT, demonstrating diffuse pathologic tracer uptake in the abdomen and further lesions in the thorax (a). Axial PET/CT view (b) shows hypermetabolic activity of omental cake, also evident in axial low-dose CT view (c). Sagittal PET/CT details (d, e) well display massive omental carcinomatosis [34].

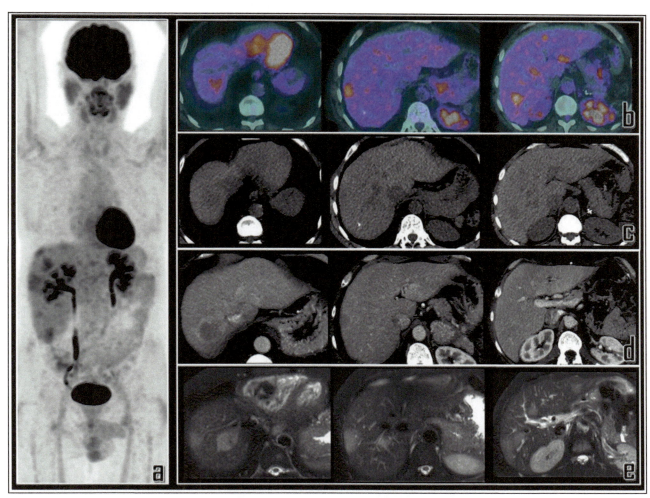

FIGURE 2.10 This figure displays the case of a 56-year-old man with history of treated *non-Hodgkin's lymphoma*, for which he had undergone chemotherapy, splenectomy and bone marrow autotransplantation. During the follow-up, this patient underwent whole body ^{18}F-FDG PET/CT and contrast-enhanced CT in single session. ^{18}F-FDG PET *Maximum Intensity Projection* (a) well displays several sites of pathologic tracer uptake in the liver, as summarized in axial PET/CT views (b). Metabolic findings are associated with hypodense lesions in correlative low-dose CT (c), with peripheral enhancement on portal venous phase in full-dose contrast-enhanced axial CT details (d). MRI, subsequently performed, confirms extranodal relapse of *non-Hodgkin's lymphoma*, underlying slightly hypointense hepatic lesions on axial T2 spectral presaturation with inversion recovery (SPIR) views (e).

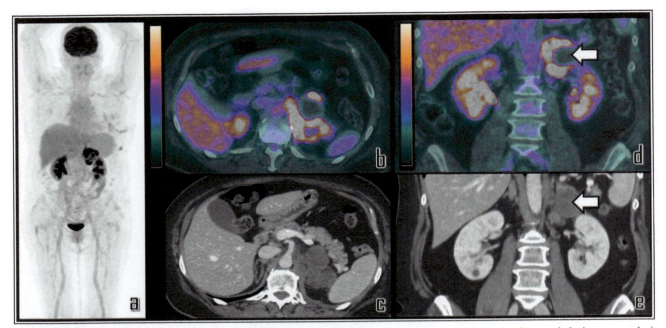

FIGURE 2.11 A patient was examined for characterization of left adrenal lesion. PET *Maximum Intensity Projection* shows pathologic tracer uptake in left adrenal gland (a), with SUVmax 9, in correspondence to a nonhomogeneous hypodense lesion, 7 cm wide, evident in axial PET/CT (b) and contrast-enhanced CT (c) views. Coronal PET/CT (d, *arrow*) and CT (e) views well display central, posthemorrhagic hypodense area, without significant ^{18}F-FDG uptake neither contrast enhancement; on the other hand, the peripheral component of the lesion presents both high gradient of glucose metabolism and meaningful contrast enhancement. Contrast-enhanced ^{18}F-FDG PET/CT for adrenal gland imaging in cancer patients allows early detection and accurate localization of adrenal lesions and differentiation of metastatic nodules from benign lesions, thereby facilitating treatment planning [35].

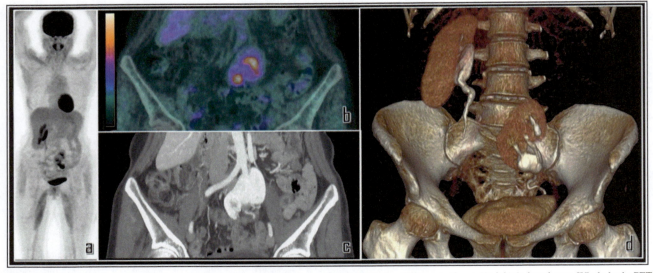

FIGURE 2.12 A 46-year-old woman was examined by contrast-enhanced PET/CT during follow-up of *non-Hodgkin's lymphoma*. Whole body PET (a) did not show disease relapse; however, the left kidney was observed in the pelvis, more eloquently showed in coronal PET/CT view (b). The ectopic kidney was also dysmorphic, with reverse rotation, as evident in correlative coronal postcontrast CT view (c) and CT volume rendering of the abdomen–pelvis (d). Renal ectopy is easily recognizable in PET/CT by identification of a kidney located below of the normal position, often mimicking a *space-occupying* lesion; the high ^{18}F-FDG uptake due to tracer excretion and correlative CT usually help in easy identification of this anatomic variant. Contrast-enhanced CT may be of help in assessing vascular features, kidney morphology, and ureteral excretion [36].

Abdomen and pelvis **Chapter** | **2** **325**

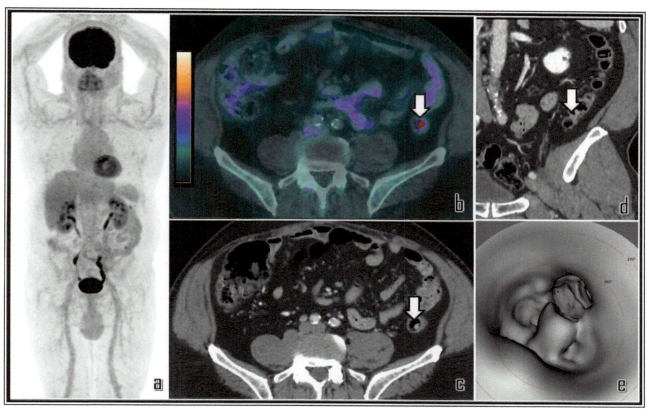

FIGURE 2.13 A 59-year-old male patient was examined by contrast-enhanced ^{18}F-FDG PET/CT during follow-up of breast cancer. ^{18}F-FDG PET *Maximum Intensity Projection* (a) only showed a focal area of mild tracer uptake (SUVmax 3) in the abdomen, in the wall of the sigma, in association with a 0.8-cm-wide node, evident in axial PET/CT (b, *arrow*) and CT (c) views and coronal CT detail (d, *arrow*). Navigation across CT colonography well displays a node protruding into the colonic lumen (e). Histological exam, performed during colonoscopy, diagnosed adenomatous polyp. Unexpected ^{18}F-FDG uptake in the intestinal loops is commonly encountered. Significant findings may be exposed by further evaluation. Endoscopy and pathology evaluation are justified, aimed to characterize these additional findings [37].

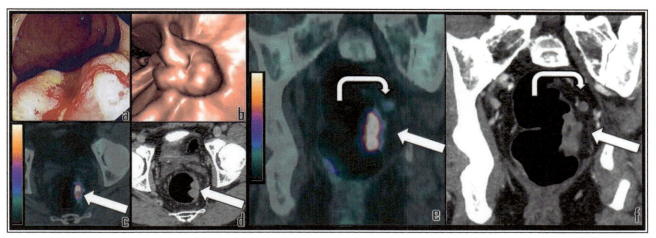

FIGURE 2.14 A 66-year-old male patient underwent melena-related colonoscopy (a) evidentiating a rectal, bleeding, and ulcerating lesion. Histological exam diagnosed colon cancer; colon exploration by colonoscopy was incomplete due to active bleeding. Therefore, patient underwent preoperative ^{18}F-FDG PET/CT integrated with CT colonography, for staging of colon cancer and complete evaluation of colic lumen in case of synchronous lesions. CT colonography confirmed a single finding in left wall of the rectum (b), evident as ^{18}F-FDG-avid, ulcerating lesion, with nonhomogeneous enhancement at contrast-enhanced PET/CT (c, d, *arrows*). Coronal PET/CT (e) and CT (f) details well display the lesion (*arrows*), also underlying a centimetric ^{18}F-FDG-avid satellite lymph node as metastatic (*curved arrows*). Usefulness of ^{18}F-FDG PET/CT integrated with CT colonography is under study, in order to improve preoperative diagnosis of obstructive colorectal cancer. Particularly, in this clinical setting, CT colonography could be of help in the assessment of synchronous tumors [38]. *Credits to Laura Travascio (MD) and Silvia Viarani (MD).*

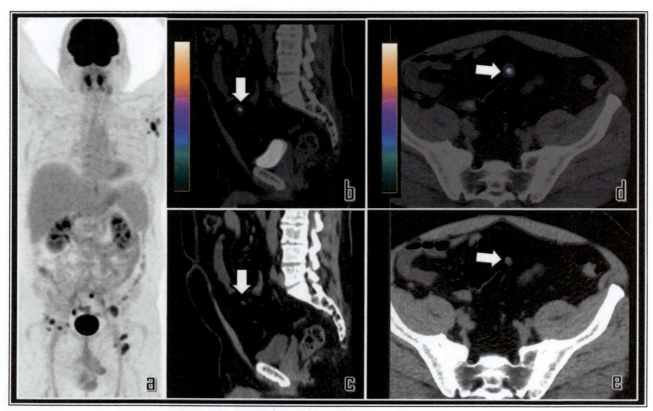

FIGURE 2.15 A 58-year-old male patient, previously submitted to chemotherapy for *non-Hodgkin's lymphoma*, underwent ^{18}F-FDG PET/CT due to palpable lymph nodes in inguinal regions and clinical suspicion of disease relapse. ^{18}F-FDG PET *Maximum Intensity Projection* (a) showed several ^{18}F-FDG-avid lymphadenopathies in the left axilla and both inguinal regions, indicative of disease relapse. A further area of focal uptake was observed in the wall of the sigma, with high SUVmax (8.9), showed in sagittal PET/CT (b) and low-dose CT (c) views (*arrows*). Axial PET/CT (d) and low-dose CT (e) views eloquently display a 9-mm wide colonic node (*arrows*). Patient underwent biopsy during colonoscopy, diagnosing colon cancer. As reported above, nuclear physicians should keep into consideration the possibility to detect abnormal sites of tracer uptake in the intestinal loops, which may be due to benign or malignant lesions. Due to low specificity of ^{18}F-FDG, focal colonic uptake should be followed up by colonoscopy [39].

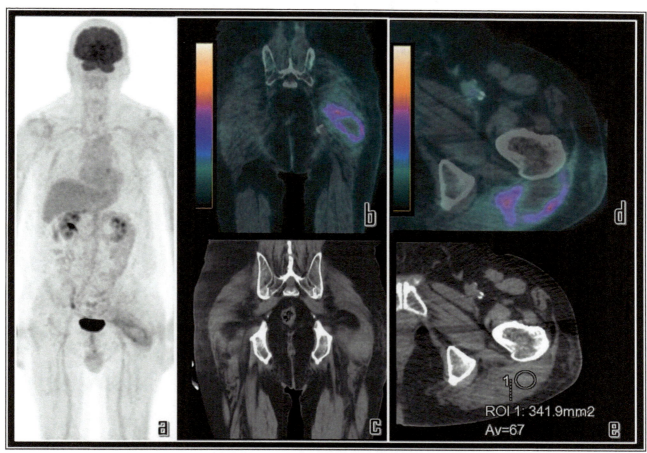

FIGURE 2.16 A 27-year-old male patient was examined by whole body ^{18}F-FDG PET/CT during follow-up of seminoma, for which he was treated with left orchifunicolectomy two years before the scan. Patient anamnesis was also positive for posttraumatic left gluteal injury. Whole body PET (a) was negative for disease relapse; however, mild and diffuse ^{18}F-FDG uptake was observed in the left gluteus, as showed in coronal PET/CT (b) and low-dose CT (c) views. Axial PET/CT (d) better displays mild ^{18}F-FDG accumulation in the left gluteus, with internal hypoenhancing center. A spheric *region of interest*, placed in the same area on correlative low-dose CT (e), also permits to assess density of 67 Hounsfield units, suggestive for red blood cells coagulation. Final diagnosis was hematoma. In this case, patient anamnesis was of the utmost importance in correct identification of a finding mimicking an occupying space lesion at PET/CT. Nevertheless, accurate knowledge of CT may improve confidence in imaging interpretation [40].

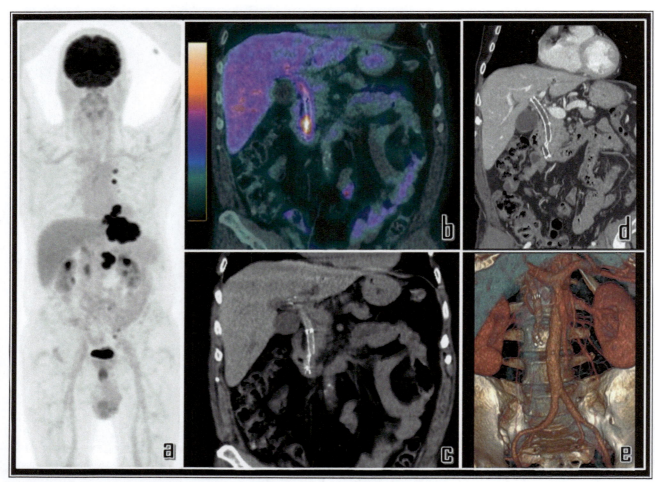

FIGURE 2.17 This 44-year-old patient was examined with contrast-enhanced ^{18}F-FDG PET/CT in order to evaluate response after second-line chemotherapy for *non-Hodgkin's lymphoma*. ^{18}F-FDG-avid supra and infradiaphragmatic lymphadenopathies were detected at whole body PET (a), indicative of absent response to treatment. Moreover, mild, diffuse tracer uptake (SUVmax 4) was observed along the tract of biliary stent, evident in coronal PET/CT (b) and low-dose CT (c) views. Correlative coronal detail of contrast-enhanced CT (d) also allowed to identify interruption and displacement of the stent in its distal tract, as also evident in 3D CT volume rendering of the abdomen (e). ^{18}F-FDG uptake may be recorded in several benign conditions not related to the disease under study [41]. Generally, these findings are related to inflammation and low SUVmax values. Contrast-enhanced CT and postprocessing of its three-dimensional views may improve diagnosis.

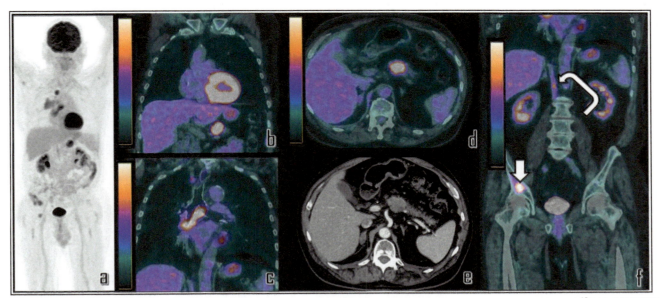

FIGURE 2.18 A 70-year-old male patient with inoperable, locally advanced nonsmall cell lung cancer, underwent contrast-enhanced ^{18}F-FDG PET/CT after chemotherapy to assess response to therapy. Whole body PET (a) showed pathologic tracer uptake in the primitive lesion in right lung and further ^{18}F-FDG-avid lesions in the mediastinum, right hemidiaphragm, right ileum, and in the abdomen. The uptake in the abdomen was linked to a lesion of the head of the pancreas, depicted in coronal PET/CT (b); further, coronal PET/CT details show pathologic subcarinal and paratracheal lymphadenopathies (c). Axial PET/CT detail better allows to recognize the lesion of the head of the pancreas, confirmed as hypodense, 3.9-cm-wide lesion with nonhomogeneous enhancement at correlative axial full-dose CT (e). Further coronal PET/CT view (f) also shows lesion of the right hemidiaphragm (*curved arrow*) and the bony localization in the right ileum (*arrow*).

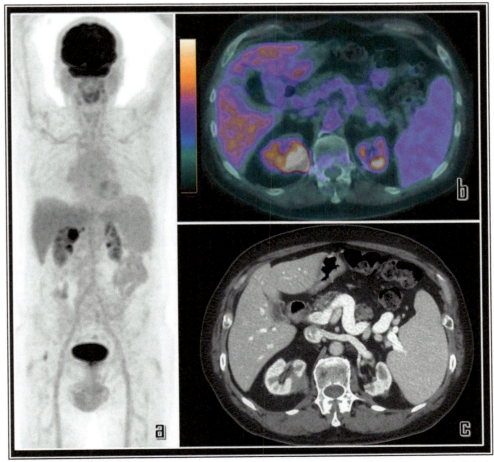

FIGURE 2.19 A 64-year-old patient previously submitted to partial left nephrectomy for renal cell carcinoma underwent whole body, contrast-enhanced ^{18}F-FDG PET/CT at follow-up. PET *Maximum Intensity Projection* (a) only showed a focal area of intense tracer uptake in the upper pole of the right kidney, higher than surrounding activity due to radiourine, more evident in axial PET/CT view (b). This metabolic finding corresponded, in correlative contrast-enhanced CT (c), to a solid tissue with irregular margins and moderate contrast enhancement, determining perinephric fat infiltration. Histological exam diagnosed renal lymphoma. It is known that renal excretion of ^{18}F-FDG [18] can lead to a misdiagnosis of primitive tumors of the kidneys. In such patient, correlative imaging with contrast-enhanced CT may help in correct identification of these findings.

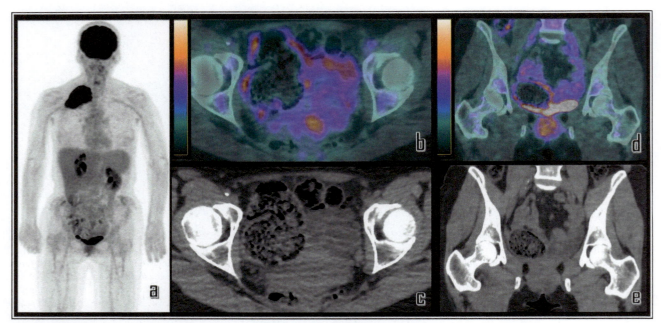

FIGURE 2.20 A 74-year-old female patient was examined by means of ^{18}F-FDG PET/CT for staging lung carcinoma. PET *Maximum Intensity Projection* (a) evidenced diffuse tracer uptake in a solid lesion of the superior lobe of the right lung, also known as *Pancoast tumor*. An area of hypometabolism was observed in the right hemipelvis, due to fecaloma, evident in axial PET/CT (b) and CT (c) views and coronal PET/CT (d) and CT (e) views. Fecal retention can be exacerbated in elderly and/or bedridden patients and occasionally can require surgical intervention [42].

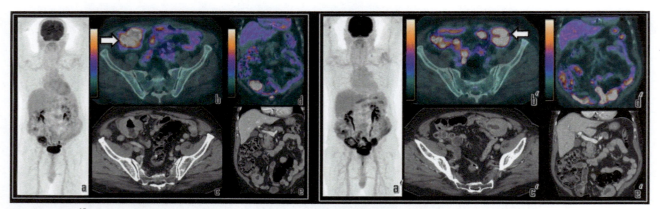

FIGURE 2.21 ^{18}F-FDG PET/CT, in a 54-year-old male patient, demonstrates tracer-avid lesion in right iliac fossa (a), close to loops of the small intestine, with high enhancement, as evident in axial PET/CT (b, arrow) and contrast-enhanced CT (c) and coronal PET/CT (d) and contrast-enhanced CT (e). Biopsy of the lesion diagnosed mantle lymphoma. Patient underwent chemotherapy. Subsequently, a further PET/CT scan demonstrated lack of response to therapy, underlying the same mass with high ^{18}F-FDG uptake in left iliac fossa (a'), more evident in correlative axial PET/CT (b', arrow) and contrast-enhanced CT (c' views) and coronal PET/CT (d') and contrast-enhanced CT (e') views. Due to physiological peristalsis, the mass followed physiological movement of intestinal loops, with a location contralateral to the origin, in the second PET/CT scan.

2.5.2 ^{68}Ga-DOTATOC

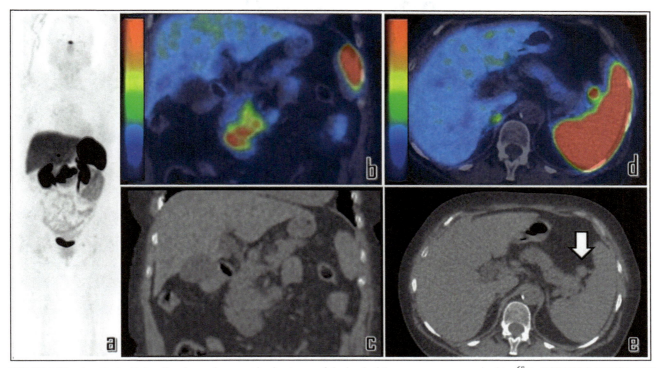

FIGURE 2.22 A patient with locally advanced neuroendocrine tumor of the head of the pancreas was examined by ^{68}Ga-DOTATOC PET/CT. PET *Maximum Intensity Projection* (a) demonstrated tracer avid lesion of the pancreas, more evident in coronal PET/CT (b) and low-dose CT (c) views. A further area of tracer uptake was detected in a 1-cm-wide, hypodense node in the adipose tissue close to the splenic hilum, showed in axial PET/CT (d) and CT (e, arrow) views, with tracer uptake gradient similar to spleen. This finding was considered as accessory spleen, a common tracer avid pitfall when using somatostatin receptor analogues [43].

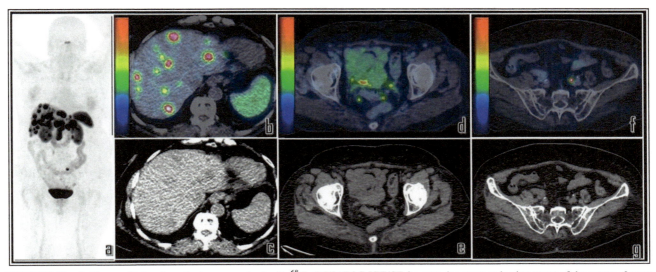

FIGURE 2.23 A 43-year-old female patient was examined with ^{68}Ga-DOTATOC PET/CT for restaging neuroendocrine tumor of the cecum, 6 years after surgery. PET *Maximum Intensity Projection* (a) displayed multiple areas of pathologic tracer uptake in the abdomen and pelvis, respectively, corresponding to several hypodense, ^{68}Ga-DOTATOC-avid hypodense hepatic lesions in axial PET/CT (b) and low-dose CT (d) views, pelvic nodes in the recto-vesical pouch, displayed in axial PET/CT (d) and low-dose CT (e) views and in a single, 0.8-cm-wide, left iliac lymph node, evident in axial PET/CT (f) and low-dose CT (g) views. Due to high expression of somatostatin receptors in neuroendocrine tumors, PET/CT with somatostatin analogues presents high sensitivity in detecting such lesions, even with the possibility to diagnose lesions not clearly pathologic at conventional radiological imaging.

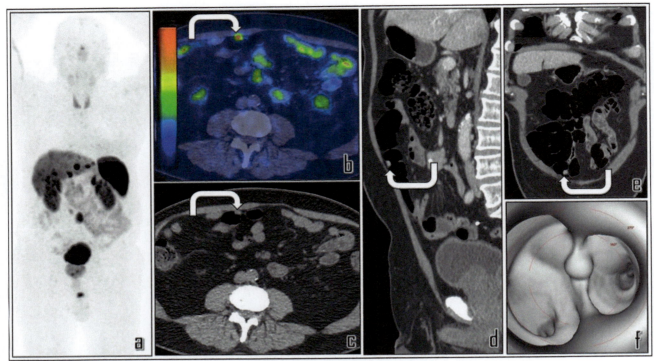

FIGURE 2.24 ^{68}Ga-DOTATOC PET/CT in a 52-year-old male patient previously submitted to surgical excision of a neuroendocrine tumor of the small intestine. PET *Maximum Intensity Projection* (a) shows several foci of pathologic ^{68}Ga-DOTATOC uptake in the liver and abdomen, due to disease relapse. A further area of focal tracer uptake was observed in a polyp of the bowel, evident in axial PET/CT (b, curved arrow) and CT (c, curved arrow) views. Integrated CT colonography allowed to better identify this 0.8-cm-wide node of the small intestine, as displayed in sagittal CT (d, curved arrow) and coronal CT (e, curved arrow) views. This finding is detectable as pedunculated during navigation in CT colonography (f) and it was considered as further localization of disease.

2.5.3 ^{18}F-Choline

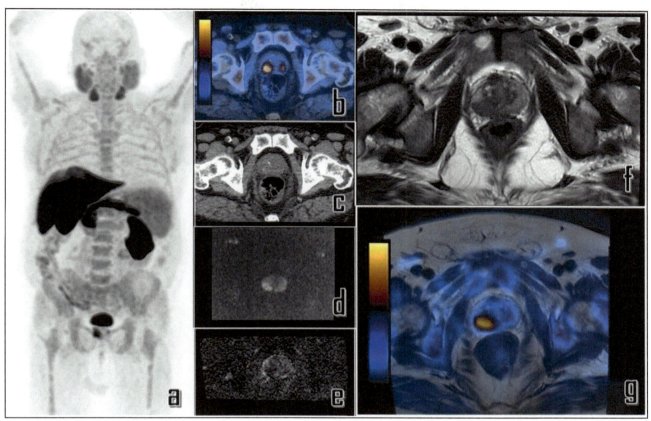

FIGURE 2.25 ^{18}F-Choline PET/CT performed in a 64-year-old patient with suspicion of prostate cancer (PSA 16 ng/mL at the time of the scan). PET *Maximum Intensity Projection* (a) shows focal tracer uptake in the pelvis, in the peripheral region of the right prostate lobe in axial PET/CT (b) and CT (c) views. Less intense tracer uptake may be also detected in left prostate lobe. No other foci of abnormal tracer uptake were detected. Correlative MRI showed hyperintensity in the same region of the right prostate lobe at DWI (d), also appearing hypointense in ADC DWI (e), confirming restricted diffusion. T2-weighted axial image (f) well displays the morphological lesion. PET/MRI view (g) obtained by postprocessing accurately shows concordance among metabolic data provided by PET and MRI findings. Histological exam diagnosed prostate cancer with Gleason 8 (4+4). MRI is the gold standard imaging tool to assess prostate cancer; in some limited cases, ^{18}F-Choline PET/CT may be of help as a guide to biopsy or to ensure distant lymph node or bony metastases [29,44].

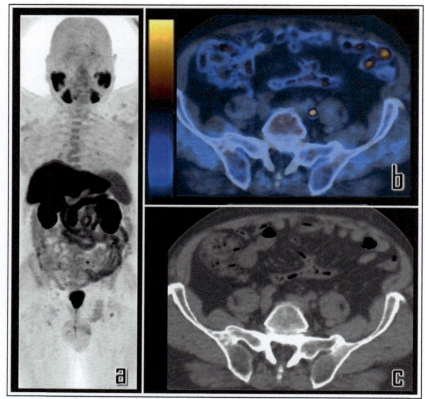

FIGURE 2.26 ^{18}F-Choline PET/CT performed in a 57-year-old patient previously submitted to radical prostatectomy for prostate cancer Gleason 7 and biochemical relapse at the time of the scan (PSA 1.3 ng/mL). PET *Maximum Intensity Projection* (a) shows a single area of intense ^{18}F-Choline uptake in a left iliac, 0.7 mm wide lymph node, site of disease relapse, showed in axial PET/CT (b) and CT (c) views. The best effort of ^{18}F-choline PET/CT is the capability to detect lymph node metastases of prostate cancer, not clearly pathologic at conventional CT imaging, ensuring early treatment in the initial phase of biochemical relapse [45].

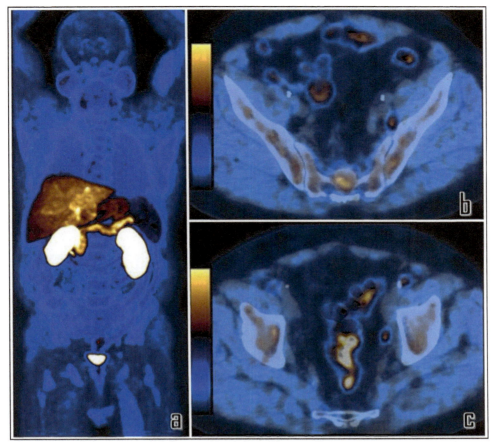

FIGURE 2.27 A patient with prostate cancer, previously submitted to radical prostatectomy, underwent ^{18}F-Choline PET/CT due to biochemical relapse (PSA 0.90 ng/ml at the time of the scan). *PSA velocity* was 1 ng/mL/year and *PSA doubling time* was less than 5 months. PET *Maximum Intensity Projection* (a) showed two areas of focal tracer uptake in the pelvis, corresponding to two subcentimeter, left iliac lymph nodes, showed in axial PET/CT views (b, c). These findings were considered as disease relapse. Derivative values of PSA kinetic as *PSA velocity* and *PSA doubling time* express the overall aggressiveness of the tumor; detection rate of 18F-choline imaging is closely related to PSA and PSA kinetics [46].

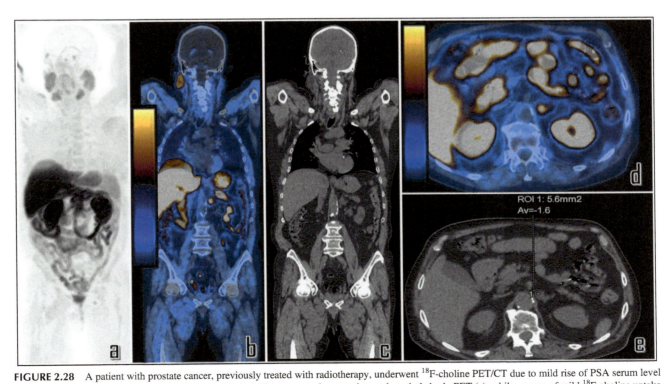

FIGURE 2.28 A patient with prostate cancer, previously treated with radiotherapy, underwent ^{18}F-choline PET/CT due to mild rise of PSA serum level (PSA 1.6 ng/ml at the time of the scan). No foci of pathologic tracer uptake were detected at whole body PET (a), while an area of mild ^{18}F-choline uptake was observed in the left adrenal gland, as evident in coronal PET/CT (b) and CT (c) views and axial PET/CT (d) view. Correlative axial low-dose CT view (e) well shows a 1-cm-wide node of the adrenal gland, suggestive for adenoma. The placed *region of interest* also permits to ensure low Hounsfield units value, confirming the diagnosis.

2.5.4 ^{68}Ga-PSMA

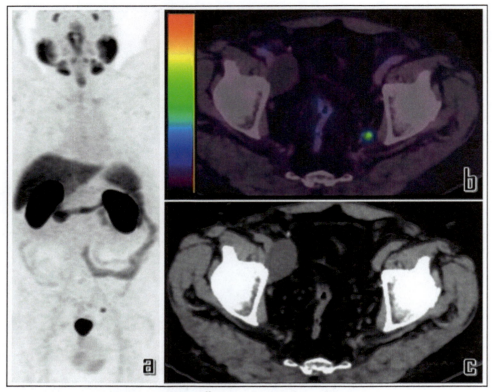

FIGURE 2.29 A 60-year-old patient was examined by means of ^{68}Ga-PSMA PET/CT for early biochemical relapse of prostate cancer (PSA 0.3 ng/mL) one year after radical prostatectomy for prostate cancer Gleason 9. PET *Maximum Intensity Projection* (a) showed single focus of pathologic tracer uptake in the left pelvis, corresponding to a 0.6-mm-wide, left iliac lymph node, indicative of disease relapse. Due to the peculiar molecular pathway, PET/CT with radiolabeled PSMA may be of help in the detection of disease relapse of prostate cancer, even with very low PSA serum level. In the axial PET/CT and CT views (b, c) is also detectable a hypodense, fluid mass in the right pelvis, recognizable as lymphocele.

2.5.5 ^{99m}Tc

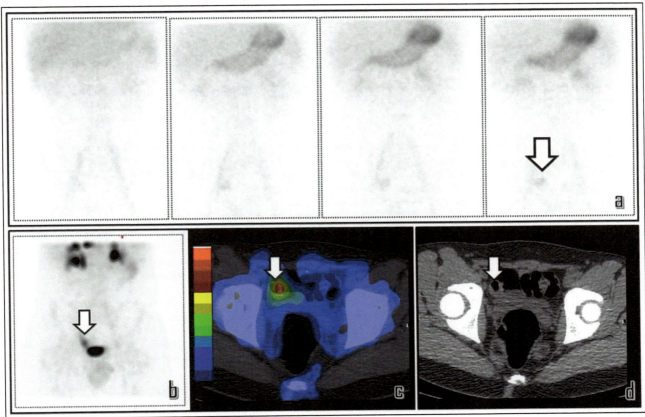

FIGURE 2.30 ^{99m}Tc-pertechnetate scintigraphy in a 12-year-old male patient with acute gastrointestinal bleeding of unknown origin and suspicion of ectopic gastric mucosa caused by Meckel's diverticulum. Dynamic, anterior planar views (a) progressively show an area of focal uptake in the right pelvis (arrow), which is visualized simultaneously with the stomach. Three-dimensional SPECT permits to better visualize this finding (b, *arrow*), while SPECT/CT (c, *arrow*) and CT (d, *arrow*) allow the exact location of the ectopic gastric tissue, adjacent to intestinal loops in right iliac fossa. Patient underwent surgical excision of this finding, confirming ectopic gastric mucosa. *Credits to Rosanna Tavolaro (MD).*

References

[1] Yamauchi FI, Ortega CD, Blasbalg R, Rocha MS, Jukemura J, Cerri GC. Multidetector CT evaluation of the postoperative pancreas. Radiographics 2012;32:743–64.
[2] Potenta SE, D'Agostino R, Sternberg KM, Tatsumi K, Perusse K. CT urography for evaluation of the ureter. Radiographics 2015;35:709–26.
[3] Fischer AM. Subcapsular collection of oral contrast material in the liver seen on CT. AJR Am J Roentgenol 2003;181:598–9.
[4] Shen Tan GJ, Berlangieri SU, Lee ST, Scott AM. FDG PET/CT in the liver: lesions mimicking malignancies. Abdom Imaging 2014;39:187–95.
[5] Keramida G, Peters AM. FDG PET/CT of the non-malignant liver in an increasingly obese world population. Clin Physiol Funct Imaging 2020;40:304–19.
[6] Nakamoto R, Okuyama C, Ishizu K, Higashi T, Takahashi M, Kusano K, et al. Diffusely decreased liver uptake on FDG PET and cancer-associated cachexia with reduced survival. Clin Nucl Med 2019;44:634–42.
[7] Halvorsen RA, Thompson WA. Computed tomography of the gastroesophageal junction. Crit Rev Diagn Imaging 1984;21:183–228.
[8] Stagg J, Farukhi I, Kaig M, Lazaga F, Gould-Simon A, Bradshaw L, et al. The significance of incidental 18F-FDG uptake in PET/CT at the gastroesophageal junction in predicting benign and malignant disease. Am J Gastroenterol 2013;108:16.
[9] Panicek DM, Benson CB, Gottlieb RH, Heitzman ER. The diaphragm: anatomic, pathologic, and radiologic considerations. Radiographics 1988;8:385–425.
[10] Keramida G, Dunford A, Kaya G, Anagnostopoulos CD, Peters AM. Hepato-splenic axis: hepatic and splenic metabolic activities are linked. Am J Nucl Med Mol Imaging 2018;8:228–38.

[11] Sugawara Y, Zasadny KR, Kison PV, Baker LH, Wahl RL. Splenic fluorodeoxyglucose uptake increased by granulocyte colony-stimulating factor therapy: PET imaging results. J Nucl Med 1999;40:1456−62.

[12] Lee JW, Kim S, Kwack SW, Kim CW, Moon TY, Lee SU. Hepatic capsular and subcapsular pathologic conditions: demonstration with CT and MR imaging. Radiographics 2008;28:1307−23.

[13] Ito K, Choji T, Fujita T, Kuramitsu T, Nakaki H, Kurokawa F, et al. Imaging of the portacaval space. AJR Am J Roentgenol 1993;161:329−34.

[14] Atay-Rosenthal S, Wahl RL, Fishman EK. PET/CT findings in gastric cancer: potential advantages and current limitations. Imaging Med. 2012;4:241−50.

[15] Blake MA, Singh A, Setty BN, Slattery J, Kalra M, Maher MM, et al. Pearls and pitfalls in interpretation of abdominal and pelvic PET-CT. Radiographics 2006;26:1335−53.

[16] Gontier E, Forume E, Wartski M, Blondet C, Bonardel G, Le Stanc E. High and typical 18F-FDG bowel uptake in patients treated with metformin. Eur J Nucl Med Mol Imaging 2008;35:95−9.

[17] Naganawa S, Yoshikawa-Shotaro T, Yasaka K, Maeda E, Hayashi N, Abe O. Role of delayed-time-point imaging during abdominal and pelvic cancer screening using FDG-PET/CT in the general population. Medicine (Baltimore) 2017;96:e8832.

[18] Lakhani A, Khan SR, Bharwani N, Stweart V, Rockall AG, Khan S, et al. FDG PET/CT pitfalls in gynecologic and genitourinary oncologic imaging. Radiographics 2017;37:577−94.

[19] Couinaud C. The Liver. Anatomical and surgical investigations. Paris: Masson. 1957. *Le foie. Etudes anatomiques et chirurgicales.*

[20] Bismuth H. Surgical anatomy and anatomical surgery of the liver. World J Surg 1982;6:3−9.

[21] Charnsangavej C, DuBrow RA, Varma DG, Herron DH, Robinson TJ, Whitley NO. CT of the mesocolon: part 1—anatomic considerations. Radiographics 1993;13:1035−45.

[22] Tirkes T, Sandrasegaran K, Patel AA, Hollar MA, Tejada JG, Tann M, et al. Peritoneal and retro- peritoneal anatomy and its relevance for cross-sectional imaging. Radiographics 2012;32:437−51.

[23] Coffin A, Boulay-Coletta I, Sebbag-Sfez D, Zins M. Radioanatomy of the retroperitoneal space. Diagn Interv Imaging 2015;96:171−86.

[24] Vesselle HJ, Miraldi FD. FDG PET of the retroperitoneum: normal anatomy, variants, pathologic conditions, and strategies to avoid diagnostic pitfalls. Radiographics 1998;18:805−23.

[25] Baseviciene I, Martinkiene I, Basevicius A, Labanauskas L. Functional ovarian cysts in girls. Medicina (Kaunas) 2003;39:902−9.

[26] Trenkner SW, Smid AA, Francis IR, Levatter R. Radiological detection and diagnosis of pouch of douglas lesions. Crit Rev Diagn Imaging 1988;28:367−81.

[27] Kim SW, Kim HC, Yang DM, Min GE. The prevesical space: anatomical review and pathological conditions. Clin Radiol 2013;68:733−40.

[28] Vargas HA, Akin O, Franiel T, Goldman DA, Udo K, Touijer KA, et al. Normal central zone of the prostate and central zone involvement by prostate cancer: clinical and MR imaging implications. Radiology 2012;262:894−902.

[29] Calabria F, Chiaravalloti A, Tavolozza M, Ragano-Caracciolo C, Schillaci O. Evaluation of extraprostatic disease in the staging of prostate cancer by F-18 choline PET/CT: can PSA and PSA density help in patient selection? Nucl Med Commun 2013;34:733−40.

[30] Morris J, Spencer JA, Ambrose NS. MR imaging classification of perianal fistulas and its implications for patient management. Radiographics 2000;20:623−35.

[31] Palmucci S, Cianci A, Ettorre GC, Basile A, Tonolini M, Foti PV, et al. Cross-sectional imaging of acute gynaecologic disorders: CT and MRI findings with differential diagnosis-part I: corpus luteum and haemorrhagic ovarian cysts, genital causes of haemoperitoneum and adnexal torsion. Insights Imaging 2019;10:119.

[32] Vaidyanathan S, Patel CN, Scarsbrook AF, Chowdhury FU. FDG PET/CT in infection and inflammation-current and emerging clinical applications. Clin Radiol 2015;70:787−800.

[33] Palomar Muñoz A, Cordero García JM, Talavera Rubio P, García Vicente AM, González García B, Bellón Guardia ME, et al. Usefulness of CA125 and its kinetic parameters and positron emission tomography/computed tomography (PET/CT) with fluorodeoxyglucose ([(18)F] FDG) in the detection of recurrent ovarian cancer. Med Clin (Barc) 2018;151:97−102.

[34] Anthony MP, Khong PL, Zhang J. Spectrum of (18)F-FDG PET/CT appearances in peritoneal disease. AJR Am J Roentgenol 2009;193:W523−529.

[35] Chong S, Lee KS, Kim HY. Integrated PET-CT for the characterization of adrenal gland lesions in cancer patients: diagnostic efficacy and interpretation pitfalls. Radiographics 2006;26:1811−24.

[36] Hadar H, Gadoth N, Gillon G. J Computed tomography of renal agenesis and ectopy. Comput Tomogr 1984;8:137−43.

[37] Goldin E, Mahamid M, Koslowsky B, Shteingart S, Dubner Y, Lalazar G, et al. Unexpected FDG-PET uptake in the gastrointestinal tract: endoscopic and histopathological correlations. World J Gastroenterol 2014;20:4377−81.

[38] Mayoral M, Rubello D, Colletti PM, Campos F, Romero I, Casanueva S, et al. PET/CT integrated with CT colonography in preoperative obstructive colorectal cancer by incomplete optical colonoscopy: a prospective study. Clin Nucl Med 2020;45:943−7.

[39] Kirchner J, Schaarschmidt BM, Kour F, Sawicki LM, Martin O, Bode J, et al. Incidental (18)F-FDG uptake in the colon: value of contrast-enhanced CT correlation with colonoscopic findings. Eur J Nucl Med Mol Imaging 2020;47:778−86.

[40] Chaber R, Łasecki M, Kwaśnicka J, Łach K, Podgajny Z, Olchowy C, et al. Hounsfield units from unenhanced ^{18}F-FDG-PET/CT are useful in evaluating supradiaphragmatic lymph nodes in children and adolescents with classical Hodgkin's lymphoma. Adv Clin Exp Med 2018;27:795−805.

[41] Shreve PD, Anzai Y, Wahl RL. Pitfalls in oncologic diagnosis with FDG PET imaging: physiologic and benign variants. Radiographics 1999;19:61−77.

[42] Plouznikoff N. Incidental detection of a giant fecaloma on 18F-FDG PET/CT. Clin Nucl Med 2020;45:83−4.

[43] Vandekerckhove E, Ameloot E, Hoorens A, De Man K, Berrevoet F, Geboes K. Intrapancreatic accessory spleen mimicking pancreatic NET: can unnecessary surgery be avoided? Acta Clin Belg 2020;12:1–4.

[44] Lovegrove CE, Matanhelia M, Randeva J, Eldred-Evans D, Tam H, Miah S. Prostate imaging features that indicate benign or malignant pathology on biopsy. Transl Androl Urol 2018;7:S420–35.

[45] Calabria F, Rubello D, Schillaci O. The optimal timing to perform 18F/11C-choline PET/CT in patients with suspicion of relapse of prostate cancer: trigger PSA versus PSA velocity and PSA doubling time. Int J Biol Markers 2014;29:e423–30.

[46] Calabria F, Tavolozza M, Caracciolo CR, Finazzi Agrò E, Miano R, Orlacchio A, et al. Influence of PSA, PSA velocity and PSA doubling time on contrast-enhanced 18F-choline PET/CT detection rate in patients with rising PSA after radical prostatectomy. Eur J Nucl Med Mol Imaging 2012;39:589–96.

Index

Note: Page numbers followed by "*f*" indicate figures.

Vol. 2

Abdominal aorta, 160, 185, 189, 191, 193, 195, 197, 199, 201, 203, 233, 249, 251, 253
Abdominal muscles, 201, 203, 205, 207, 209, 211, 213, 215, 227, 241, 243
Abdominal topography, 159f
Adductor muscle, 225
Adenoma, 147f, 320f, 325f, 335f
Adrenal gland, 324f, 335f
Adrenal lesion, 324f
Anal canal, 221, 223, 235, 253, 303
Anal clock, 314f
Anterior basal segment of left lung [S8, inferior lobe], 39, 41, 43, 45, 47, 49, 69, 71, 73, 75, 77, 79, 81, 83, 85, 87, 89
Anterior basal segment of right lung [S8, inferior lobe], 39, 41, 43, 45, 47, 57, 59, 61, 77, 79, 81, 83, 85, 87, 89, 91
Anterior renal fascia, 286f, 288f, 290f, 292f, 294f
Anterior segment of left lung [S3, superior lobe], 15, 17, 19, 21, 23, 25, 27, 29, 31, 65, 67, 69, 71, 73, 75, 77, 79, 81, 83, 85, 87, 89
Anterior segment of right lung [S3, superior lobe], 15, 17, 17f, 19, 21, 23, 25, 27, 29, 31, 55, 57, 59, 61, 63, 75, 77, 79, 81, 83, 85, 87, 89
Anus, 225, 255, 313
Aorta bifurcation («carrefour»), 205
Aortic arch, 2f, 107
Aortic bulb, 115
Aortic hiatus, 188f
Apical segment of right lung [S1, superior lobe], 11, 13, 15, 17, 19, 21, 23, 55, 57, 59, 61, 63, 75, 77, 79, 81, 83, 85, 87, 89, 91
Apicoposterior segment of left lung [S1+2, superior lobe], 11, 13, 15, 17, 19, 21, 23, 25, 27, 29, 65, 67, 69, 71, 73, 75, 77, 79, 83, 85, 87, 89, 91, 93
Ascending aorta, 5, 109, 111, 113
Ascending colon, 167, 197, 197f, 199, 201, 203, 205, 241, 245, 247, 249, 279, 281, 291
Atelectasia, 133f
Axilla, 109, 111, 140f, 326f
Azygos vein, 121, 123, 125, 127, 128f, 175

Bare area of liver, 293f
Barety's space, 145f
Bifid ureter, 317f
Biochemical relapse, 147f—150f, 334f—336f
Bismuth classification, 263f
Bone metastases, 150f
Brachiocephalic artery, 105
Brachiocephalic vein, 101, 103, 105
Bronchioloalveolar carcinoma, 131f

Carina of trachea, 3, 27
Carotid artery, 101, 103
Caudate lobe, 179, 181, 261f
Cecum, 167, 207, 209, 331f
Celiac trunk, 185
Cervix, 301, 321f
Clavicle, 97, 99, 101, 103, 105, 107, 150f
Colic flexure, 167
Colon, 134f, 142f, 183, 185, 187, 189, 191, 193, 195, 197, 199, 201, 203, 205, 207, 209, 211, 213, 217, 227, 229, 231, 233, 235, 237, 241, 243, 245, 247, 249, 273, 279, 281, 283, 285, 287, 289, 291, 293, 295, 297, 325f
Colon cancer, 134f, 142f, 325f—326f
Colorectal cancer, 325f
Common iliac artery and vein, 207, 209, 211, 233, 237
Coronary sinus, 121
Corpus cavernosum, 313, 315
Corpus luteum, 318f
Corpus spongiosum, 315
Costochondral joint, 109, 111
Costovertebral joint, 123, 125
Couinaud classification, 256f
Crura, 179f, 237
Crus, 179, 235

Deltoid, 97, 99, 101, 103, 105
Descending aorta, 109, 111, 113, 115, 117, 119, 121, 123, 125, 127
Diaphragm, 95f, 125, 179, 181, 235, 237
Dorsal vertebra, 97, 99, 101, 103, 107, 109, 113, 115, 117, 119, 121, 123, 125
Ductus deferens, 309f, 315
Duodenal bulb, 187
Duodenal flexure, 163
Duodenum, 163, 189, 191, 193, 195, 197, 199, 201, 233, 235, 237, 247, 249, 320f

Elastofibromadorsi, 142f
Emiazygos vein, 125, 127, 175
Epigastric region, 158f
Erector spinae, 97, 99, 101, 103, 109, 115, 117, 121, 123, 125, 127
Esophagus, 97, 99, 101, 103, 105, 107, 109, 111, 113, 115, 121, 123, 125, 171, 173, 175, 177, 177f, 233, 253
External iliac artery and vein, 213, 215, 217, 219
External oblique muscle, 123, 125, 127

Falciform ligament, 257f, 265, 267, 275, 275f, 277
Female pelvis, 296f
Femoral artery and vein, 221, 223, 225
Fissures (obliques/horizontal), 3

Gallbladder, 163, 183, 185, 187, 189, 239, 243, 245, 265, 267, 267f, 269, 281, 293
Gallbladder fossa, 265, 267, 267f
Gastro-intestinal stromal tumor (Gist), 319f
Gastroesophageal junction,, 177, 177f, 233
Gastrohepatic (hepatogastric) ligament, 181
Gated PET/CT, 129f
Glisson's capsule, 171, 171f, 173, 183f
Gluteus muscle, 207, 209, 211, 213, 215, 217, 219, 221, 223, 225, 255
Greater curvature of stomach, 163

Heart, 5, 95f, 171, 173
Hematoma, 327f
Hemiazygos vein, 125, 127, 175
Hepatic vein, 161, 173, 175, 257, 259, 261, 263
Hepato-splenic axis, 181f
Hepatorenal recess (subhepatic recess), 271
Hernia, 318f
Hodgkin's lymphoma, 115f, 138f, 318f
Horizontal fissure of right lung, 9
Horseshoe kidney, 128f
Hypocondriac region, 158f
Hypogastric (pubic) region, 158f

Ichial tuberosity, 311
Ileal and jejunal arteries and veins, 201, 203
Ileum, 146f, 203, 205, 207, 209, 211, 213, 237, 239, 243, 245, 247, 293, 329f

341

342 Index

Iliacus muscle, 207, 209, 211, 213, 213f, 231, 251, 253, 297, 299
Iliopsoas muscle, 213f, 215, 217, 219, 221, 223, 225, 229, 241, 247
Inferior lingular segment of left lung [S5, superior lobe], 33, 35, 37, 39, 41, 43, 45, 49, 53, 197, 203, 205, 207
Inferior lobar bronchus, 5, 35, 37
Inferior vena cava, 121, 160—161, 171, 173, 175, 177, 179, 181, 183, 185, 185f, 187, 189, 191, 193, 195, 197, 199, 201, 203, 205, 235, 237, 251, 257, 257f, 259, 261, 263, 265, 267, 269, 271, 279
Inflammation, 132f, 148f—149f, 328f
Inframesocolic peritoneal spaces, 279
Infraspinous muscle, 107, 115
Inguinal region, 326f, 158f
Intercostal muscle, 107, 109, 115, 119, 121, 125, 127
Intergluteal cleft, 313
Interim PET, 115f, 138f
Intermediate bronchus, 5, 31, 33, 35
Internal carotid, 97, 99
Internal iliac artery and vein, 213, 215
Intertubercular plane, 158f
Intravenous stasis, 140f
Ischioanal fossa, 303, 305, 313
Ischiocavernosus muscle, 235, 313
Ischiorectal fossa, 301, 309, 311

Jejunum, 189, 191, 193, 195, 197, 199, 201, 203, 205, 207, 209, 227, 229, 231, 233, 235, 243, 245, 247, 251, 273, 287, 289, 291, 295

Kidney, 160—161, 183, 185, 187, 189, 191, 193, 195, 197, 199, 201, 207f, 229, 231, 239, 241, 251, 253, 255, 281, 285, 287f, 316f, 324f, 329f

Lateral basal segment of left lung [S9, inferior lobe], 39, 41, 43, 45, 47, 49, 51, 69, 71, 73, 85, 87, 89, 91, 93
Lateral basal segment of right lung [S9, inferior lobe], 41, 43, 45, 47, 49, 55, 57, 59, 75, 79, 81, 83, 85, 87, 89, 91, 93
Lateral (lumbar) region, 128f, 321f, 158f
Lateral segment of right lung [S4, middle lobe], 29, 31, 33, 35, 37, 39, 41, 43, 55, 57, 77, 79, 81, 83, 85, 87, 89, 91
Lateroconal fascia, 291f
Latissimusdorsi muscle, 179, 227, 239, 241
Left atrium, 115, 117, 119
Left auricle, 113
Left hepatic lobe, 125, 127
Left ventricle, 115, 117, 119, 121, 123
Lesser curvature of stomach, 163
Lesser sac or omental bursa, 273
Levator platform, 313f
Ligamentumvenosum, 179
Lingular segmental bronchus of left lung, 33, 35
Liposarcoma, 139f

Liver, 63f, 121, 123, 123f, 125f, 160, 171f, 173f, 181f, 183f, 195, 243, 245, 247, 249, 251, 253, 255, 257f, 261f, 265, 265f, 267f, 273, 275, 275f, 277, 281, 293, 293f, 332f
Lobar bronchus, 5, 27, 31, 35, 37, 181
Longissimusdorsi muscle, 197
Lung, 15, 17, 19, 197, 203
Lung cancer, 47f, 179f, 316f, 319f—320f, 325f, 329f, 337f
Lung nodule, 47f
Lymphocele, 336f

Main bronchus, 3, 27, 29, 31, 33, 111
Male pelvis, 309f
Mantle lymphoma, 330f
Manubriosternal joint, 111
Meckel's diverticulum, 337f
Medial basal segment of left lung [S7, inferior lobe], 43, 45, 47, 67, 85, 87, 89
Medial basal segment of right lung [S7, inferior lobe], 41, 43, 75, 77, 79, 81, 83
Medial segment of right lung [S5, middle lobe], 33, 35, 37, 39, 41, 43, 45, 47, 57, 59, 61, 63, 77, 79, 81, 83
Median plane, 159f
Mediastinal compartments, 133f, 145f
Mediastinum, 316f, 324f, 329f, 331f—332f, 252f
Medullary thyroid cancer, 145f
Melanoma, 135f—136f
Mesorectal fascia, 309f
Midclavicular plane, 158f
Middle lobar bronchus, 5, 37
Morison's pouch, 273, 316f

Neuroendocrine tumor, 146f, 331f—332f
Nipple, 117
Non-Hodgkin Lymphoma, 143f

Oblique fissure, 9
Obturator canal, 301, 303
Obturatorexternus muscle, 219, 221, 223, 305
Obturatorinternus muscle, 217, 219, 221, 223, 299, 301, 303, 305, 309, 311
Omental bursa (lesser sac), 163
Omental cake, 322f
Omentalcarcinomatosis, 322f
Osteophitosis, 153f
Ovarian cancer, 320f, 322f
Ovary, 297, 299

Pacemaker, 143f
Pampiniform plexus, 315
Pancoasttumor, 330f
Pancreas, 163, 164f—165f, 183, 185, 187, 189, 191, 193, 195, 197, 229, 231, 233, 235, 249, 251, 277, 283, 283f, 289f, 329f, 331f
Paracolic space, 279
Parathyroid adenoma, 147f
Parenchymal consolidation, 334f—335f

Parietal peritoneum, 287f, 289f, 291f
Pectineus muscle, 223, 303
Pectoralis major, 97, 99, 101, 103, 105, 107, 111, 113, 115, 117, 119
Pectoralis minor, 99, 105, 107, 109, 111
Pelvis, 137f, 148f, 150f, 291f, 331f
Perianal fistula, 319f
Perianal region, 315f
Perineum, 132f, 296f
Perirectal space, 309f
Perirenal space, 287f, 291f, 293f
Perisplenic space, 273
Peritoneal carcinomatosis, 320f
Peritoneum, 247f, 273f, 287f, 289f, 291f
Piriformis muscle, 219, 221, 297, 299
Pleural carcinomatosis, 141f
Pneumonitis, 136f, 149f
Polyp, 325f, 332f
Portacaval space, 185, 185f
Portal vein, 165f, 185f, 263, 263f, 265, 265f, 267
Posterior basal segment of left lung [S10, inferior lobe], 39, 41, 45, 47, 49, 51, 53, 65, 67, 69, 71, 73, 85, 87, 89, 91, 93
Posterior basal segment of right lung [S10, inferior lobe], 39, 49, 53, 85, 87, 89, 91, 93
Posterior border of the lung, 3
Posterior renal fascia, 287f, 289f, 291f, 293f, 295f
Posterior segment of right lung [S2, superior lobe], 15, 17, 17f, 19, 21, 23, 25, 27, 55, 57, 59, 61, 63, 85, 87, 89, 91, 93
Prostate, 147f—149f, 309, 334f
Prostate cancer, 147f—150f, 153f, 311f, 333f—336f
Psoas major muscle, 211, 213, 213f, 231, 237, 239, 251, 253, 255
Pubic simphysis, 221, 223
Pubococcygeus muscle, 309
Puborectalis muscle, 301, 303, 305, 309, 311
Pulmonary artery, 109, 111
Pulmonary trunk, 5
Pulmonary vein, 5
Pyloric canal, 163
Pyloric pars, 163
Pylorus, 163

Quadratusfemoris muscle, 223, 225
Quadratuslumborum muscles, 201, 203, 205, 227, 229, 231, 239, 241

Radiotherapy, 131f, 133f, 144f, 149f, 322f, 335f
Radiourine activity, 207f
Rectouterine pouch (of douglas), 299
Rectovesical pouch, 279—285
Rectum, 167, 217, 219, 235, 253, 255, 283, 299, 301, 325f
Rectus abdominis muscle, 127, 217, 229, 233, 235
Rectus femoris muscle, 219, 221, 223, 225
Renal artery, 195
Renal cell carcinoma, 329f

Renal ectopy, 324f
Renal pelvis, 193, 197
Renal vein, 161, 193, 195
Respiratory gating, 47f
Response to therapy, 115f, 137f–138f, 142f, 144f, 329f–330f
Retrocrural space, 181, 181f, 183
Retroperitoneum, 273f, 287f, 289f, 291f, 293f, 295f
Retzius, 301, 301f, 303
Rib, 97, 99, 107, 111, 113, 115, 117, 119, 123, 125, 127, 145f–146f, 150f, 152f
Rib metastases, 150f
Right atrium, 115, 117, 119
Right auricle, 5, 113
Right pulmonary artery, 111
Right ventricle, 113, 115, 117, 119, 121, 123
Root penis, 225
Round ligament, 181, 183, 185

Sarcoidosis, 137f
Sartorius muscle, 215, 217, 219, 221, 223, 225
Scapula, 97, 99, 101, 103, 105, 107, 109, 113, 115, 117, 119, 121
Segmental anatomy, 35
Seminal vesicle, 309f
Seminoma, 144f, 327f
Serratus anterior muscle, 117, 119, 121, 123
Sigmoid colon, 167, 297, 299
Sine materia uptake, 134f
Small cell lung cancer, 329f
Solitary lung metastases, 142f
Solitary lung node, 129f–131f, 146f
Spinal cord, 97, 99, 101, 103, 105, 107, 109, 111, 117, 119, 121, 123, 125, 127
Spinalis muscle, 197, 211
Spine, 95f, 97, 107, 113, 117, 119, 123, 287f

Spinocellular carcinoma, 150f
Spleen, 160, 273, 279, 285, 295
Splenic hilum, 160, 165, 331f
Splenic vein, 160, 187, 269
Spleno-mesenteric venous confluence, 165f
Spleno-renal recess, 161
Sternoclavicular joint, 105, 143f
Sternocleidomastoid, 97, 99
Sternocostal joint, 125
Sternum, 95f, 105, 107, 109, 113, 115, 117, 119, 121, 123, 125
Stomach, 127, 163, 185, 187, 189, 191, 193
Subclavian artery, 101, 103, 105
Subclavian vein, 101, 103, 105
Subclavius muscle, 97, 99
Subcostal plane, 158f
Subcutaneous adipose tissue, 101, 103, 119, 121
Subhepatic space, 273
Subscapularis muscle, 97, 99, 101, 103, 105, 107, 113, 115
Superior lingular segment [S4, superior lobe], 35
Superior mesenteric artery, 191, 193
Superior mesenteric vein, 160, 164f, 189, 191, 193, 195, 197, 199
Superior segment of left lung [S6, inferior lobe], 23, 25, 27, 29, 31, 33, 35, 37, 65, 67, 69, 71, 73, 91, 93
Superior segment of right lung [S6, inferior lobe], 23, 25, 27, 29, 31, 33, 35, 37, 55, 57, 59, 61, 63, 91, 93
Superior vena cava, 5, 107, 109, 111, 113
Supramesocolic peritoneal spaces, 273
Supraspinatus muscle, 97, 99, 101

Tensor fasciae latae muscle, 217, 219, 221, 223, 225

Thoracic aorta, 171, 173, 175, 177, 179, 181, 183, 187, 189f
Thymic activity, 128f
Thyroid cancer, 145f, 151f
Thyroid goiter, 148f
Trachea, 3, 11, 13, 13f, 17, 17f, 19, 21, 23, 25, 27, 97, 99, 101, 103, 105, 107, 109, 132f
Tracheobronchitis, 132f
Transumbilical plane, 159f
Transversalis fascia, 287f, 289f, 291f, 293f, 295f
Transverse colon, 167
Transverse fissure, 181
Trapezius muscle, 97, 99, 101, 103, 107, 109, 111, 113, 115, 117
Triangular ligament, 275

Umbilical (mesogastric) region, 159f, 158f
Uncinate process, 164f, 165, 191, 193, 195, 197
Ureter, 168–169, 168f, 317f
Ureteral cross, 168
Ureteropelvic junction, 168–169
Ureterovesicular junction, 169
Urethra, 305, 309f, 311
Urinary bladder, 219, 219f, 233, 235, 237, 247, 249, 251
Uterine cervical cancer, 321f
Uterus, 297, 299, 301

Vagina, 303, 305, 321f
Vaginal fornix, 321f
Vascular central compartment, 287f
Vertebra, 243, 249, 251, 253, 255, 263

Winslow, 277f

Xiphoid process, 177, 275